Algebra 1
Principles of
Secondary Mathematics

Student Worktext
Book A

888-854-6284
mathusee.com
customerservice@demmelearning.com

Algebra 1: Principles of Secondary Mathematics Student Worktext, Books A and B
©2022 Demme Learning, Inc.
Published and distributed by Demme Learning

All rights reserved. No part of this book may be reproduced, stored in a retrieval system, or transmitted in any form by any means—electronic, mechanical, photocopying, recording, or otherwise—without prior written permission from Demme Learning.

mathusee.com

1-888-854-6284 or 1-717-283-1448 | demmelearning.com
Lancaster, Pennsylvania USA

ISBN 978-1-60826-215-1 (Algebra 1 Student Worktext)
ISBN 978-1-60826-213-7 (Book A)
Revision Code 0522

Printed in the United States of America by The P.A. Hutchison Company
3 4 5 6 7 8 9 10

For information regarding CPSIA on this printed material call: 1-888-854-6284
and provide reference # 0522-10132022

Table of Contents

This book includes all content listed for Book A.

Student Worktext Book A

Get Started with Algebra 1

Welcome...v
Course Structure......................................v
Student Components................................vi
Student Role...vii
Plan, Implement, Explain Method..............vii
Quick Start Guide.................................viii
How to Complete a Lesson.......................ix

Unit 1: Foundations of Algebra

Record Keeping and Objectives...................1
Lesson 1: The Language of Algebra..............3
Lesson 2: Solving Equations.......................23
Lesson 3: Solving Absolute Value Equations...43
Lesson 4: Solving Inequalities....................61
Lesson 5: Ratios, Proportions, and Rates......79
Lesson 6: Understanding Data....................95

Unit 2: Linear Equations and Functions

Record Keeping and Objectives.................125
Lesson 7: Functions................................127
Lesson 8: Using Graphs...........................147
Lesson 9: Slope and Linear Equations.........165
Lesson 10: Writing Linear Equations..........189
Lesson 11: More Forms of Lines................209
Lesson 12: Parallel and Perpendicular Lines...227
Lesson 13: Scatter Plots..........................251
Lesson 14: Types of Functions and Arithmetic Sequences.............273

Unit 3: Systems of Equations and Inequalities

Record Keeping and Objectives.................293
Lesson 15: Graphing Systems of Linear Equations and Inequalities.......295
Lesson 16: Solving Systems of Equations Algebraically...............315
Lesson 17: Applications of Linear Systems...335
Lesson 18: More Applications of Linear Systems..353

Index..371

Student Worktext Book B

Welcome to Book B

Welcome..vii
Course Structure...................................vii
Quick Start Guide................................viii

Unit 4: Polynomial Expressions and Equations

Record Keeping and Objectives...................1
Lesson 19: Product and Power Rules for Exponents.......................3
Lesson 20: Polynomial Expressions..............19
Lesson 21: Introduction to Factoring Polynomials...39
Lesson 22: Patterns in Factoring................53
Lesson 23: Factoring Polynomials...............69
Lesson 24: More Factoring........................85
Lesson 25: Applications of Quadratics..........99

Unit 5: Quadratics, Exponentials, Radicals

Record Keeping and Objectives.................117
Lesson 26: Graphing Quadratics................119
Lesson 27: More Quadratic Graphing..........141
Lesson 28: More Exponent Rules...............163
Lesson 29: Radical Expressions and Equations....181
Lesson 30: Exponential Functions..............201

Index..225

Digital Pack

The Digital Pack contains essential lesson videos and resources as noted in the Quick Start Guide. Log in at digital.demmelearning.com.

Welcome to Algebra 1!

Hi! I'm Sara Donovan, the curriculum writer and instructor of this course. I have been a middle and high school math teacher for nearly two decades, and I am excited to share my love of math and learning with you.

In this curriculum, you will learn about the following topics:

- solving equations and inequalities
- writing and graphing linear equations
- solving systems of equations and inequalities
- factoring polynomials
- graphing nonlinear equations

All of these topics work together to build a foundation for secondary math.

Course Structure

Algebra 1: Principles of Secondary Mathematics has five units. Each unit contains several lessons and a unit test. Each lesson has a Part A and Part B, a Lesson Test, and a Targeted Review. The course also includes a Midterm Exam and a Final Exam.

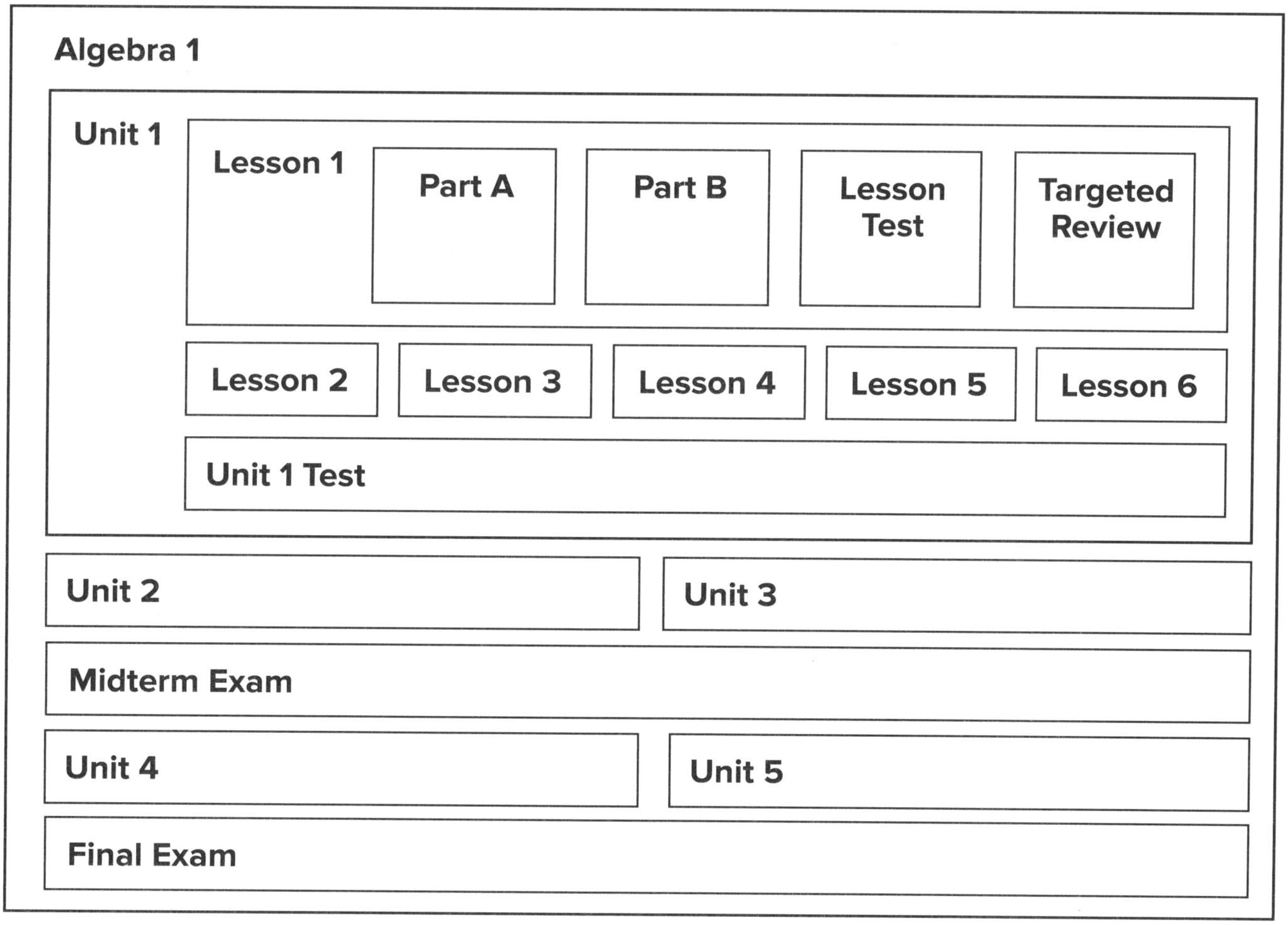

Algebra 1

- **Unit 1**
  - Lesson 1: Part A | Part B | Lesson Test | Targeted Review
  - Lesson 2 | Lesson 3 | Lesson 4 | Lesson 5 | Lesson 6
  - Unit 1 Test
- **Unit 2** | **Unit 3**
- **Midterm Exam**
- **Unit 4** | **Unit 5**
- **Final Exam**

GET STARTED WITH ALGEBRA 1

Student Components

Most of your time in this course will be spent working in your Student Worktext and watching videos in the Digital Pack. Here are all the components you will use:

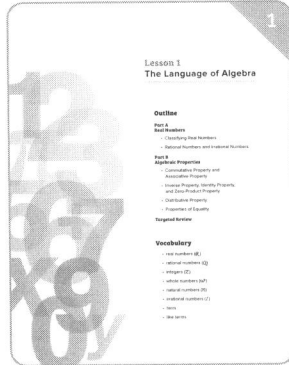

Student Worktext

Each lesson is divided into Part A and Part B. The sections described here are found in both parts.

- Lesson Objectives
    - Mastery of objectives leads to test readiness
- Warm Up
    - Activities to prepare your mind for that part of the lesson
    - Five minutes or less to complete
- Explore

    Within Explore, you may encounter multiple cycles of video, notes, and a Checkpoint. Every Explore section includes
    - Cues to watch videos in Digital Pack
    - Guided notes and examples
        - Demonstrated in videos
        - Completed in Student Worktext
    - Checkpoint problems to verify understanding
- Practice 1 and Practice 2 to build, practice, and reinforce explored concepts
- Mastery Check to gauge mastery of concepts and test preparedness
- Targeted Reviews (only after Part B) for gradual Unit Test preparation

Digital Pack

- Video presentations of concepts, notes, and examples with a math instructor
- Worked solutions for Practices, tests, exams, and Targeted Reviews
- Digital manipulatives for visualizing problems and solutions in the course
- Additional resources including links to Desmos®, graph paper, extension lessons, etc.

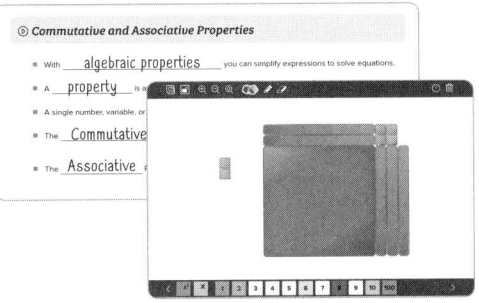

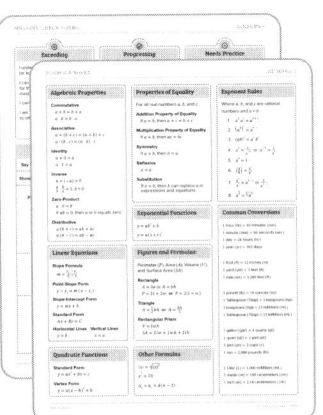

Formula Sheet/Mastery Rubric

- Formulas, algebraic properties, and conversions needed for the course
- Formula Sheet use is encouraged throughout the course
- Copy of a rubric to be used for self-evaluation and/or instructor evaluation of Mastery Check, individual problems, lessons, units, etc.

Tests

- Lesson and Unit Tests
- Midterm Exam and Final Exam
- Answer Keys for all Tests and Exams

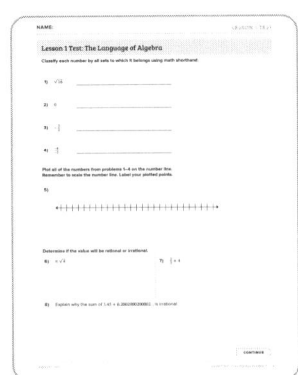

Student Role

There are a few places that you as the student should check in with your instructor to make sure you are moving through the curriculum successfully. You should check in with your instructor at these times:

- When you complete a Checkpoint
- When you complete Practice 1
- When you complete the Show What You Know part of the Mastery Check
- When you are ready to do the Say What You Know part of the Mastery Check
- When you complete Practice 2
- When you complete the Targeted Review
- Any other place in a lesson where you get stuck or need help

Plan, Implement, Explain Method

In Algebra 1, you will learn to use the Plan, Implement, Explain method for problem-solving.

Plan Plan how you will approach the problem.
- Examine the problem.
  - What is the problem asking you to do?
  - What information do you need to complete the problem?
- Create a plan to find a solution.
- Determine which math strategies apply.
- Decide what tools can be used.

Implement Implement your plan to complete the problem, then check your work.

Implement
- Use the plan to complete the problem.
- Show your work.
- Label important information.

Check
- Verify the answer is complete.
- Make sure the answer makes sense.
- Substitute the answer into the problem using a calculator as needed.
- If your answer does not make sense, revisit your plan and try a different strategy.

Explain Explain why your answer makes sense for the given problem.
- Explain why your answer makes sense.
- Connect previous knowledge to new concepts.
- Name any mathematical properties used.
- For word problems, write answers in a complete sentence.

The implementation and explanation of problem-solving work together. As you work through a problem, you should be able to explain each step as you implement it.

There are times when you will be asked to document all parts of this method. Other times, you may complete Plan and/or Explain in your mind. Always read the directions for each problem carefully to see which parts of the method you must show.

GET STARTED WITH ALGEBRA 1

Quick Start Guide

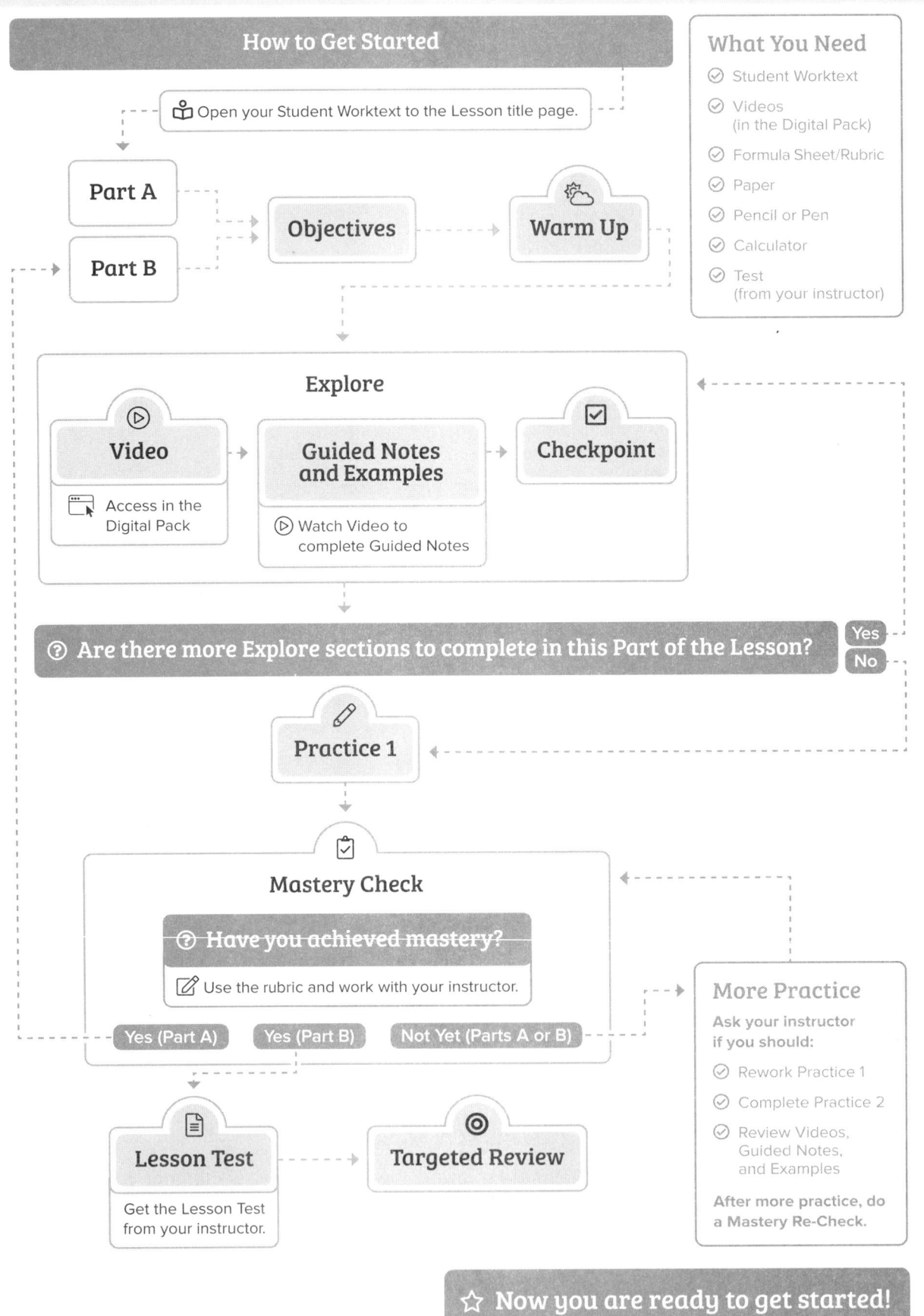

How to Complete a Lesson

Each lesson is divided into three sections, Part A, Part B, and the Targeted Review. There is also a Lesson Test. This section describes how to complete a lesson.

Part A

- As you begin a lesson, you should read the **Objectives** and the **Why** statement to get a feel for what you will be learning.
- Complete the **Warm Up**. The Warm Up often includes a review of something that is related to or important to your understanding of that part of the lesson. It is a great way to get you focused on what you are about to learn.
- Next you will encounter the **Explore** section. Although there is no Explore title, you can find it either by looking in the page header, or by recognizing that Explore has begun when you see the first video icon. The Explore sections include videos, guided notes, examples, and a Checkpoint.
    - When you see a video icon and a section title, that is your cue to start the **video** with the same title in the Digital Pack.
    - As you watch the video, fill in the **guided notes** and the missing parts of the **examples** in your Student Worktext as they are revealed.
    - If the video is going too fast for you, you can pause it. If you missed something, you can back up and watch that part again.
    - When you have completed the video and the guided notes and examples, you will encounter a Checkpoint. A **Checkpoint** is a problem that you can complete on your own based on what you have learned through the video.
- After you have completed all of the Explore section, you will find **Practice 1**. This is an opportunity for you to practice on your own what you have learned in this part of the lesson. You should complete practice problems on a separate sheet of paper so that you have plenty of room to work through them.
- After Practice 1, you have a **Mastery Check**. The Mastery Check is a problem or a series of problems and questions that help determine if you have mastered the concepts. You can use the provided rubric to self-evaluate your work. This is the same rubric that your instructor will use to confirm mastery.
- Based on the outcome of your Mastery Check, you and your instructor can decide whether or not you should re-work any Practice 1 problems, complete **Practice 2**, or do any other review of Part A before moving on to Part B.

Part B

You will work through Part B of a lesson the same as you did for Part A.

Lesson Test

After mastering the concepts in Parts A and B, you and your instructor can decide if you are ready to take the Lesson Test.

Targeted Review

The Targeted Review is designed to help you build and maintain your mastery of concepts. These reviews combine problems from previous lessons and courses. It is recommended that you complete the Targeted Review after taking the Lesson Test.

UNIT 1   Record Keeping   Name: Algebra 1

Lesson		Part	Guided Notes	Practice 1	Practice 2	Mastery Check	Lesson Test	Targeted Review
1	The Language of Algebra	A Real Numbers						
		B Algebraic Properties						
2	Solving Equations	A Multi-Step Equations						
		B Equations with Special Cases						
3	Solving Absolute Value Equations	A Absolute Value Equations						
		B Multi-Step Absolute Value Equations						

Unit Test I   Date _____   Score _____

Lesson Objectives

Lesson 1: Part A
- ☐ Identify numbers as real, rational, integers, whole, natural, and/or irrational.
- ☐ Draw a scaled number line showing numbers and approximate numbers.
- ☐ Explain why
  - a rational number + a rational number = a rational number
  - a rational number + an irrational number = an irrational number
  - an irrational number · an irrational number = an irrational number

Lesson 1: Part B
- ☐ Determine when an absolute value equation has no solution or is an identity.

Lesson 2: Part A
- ☐ Identify algebraic properties within an equation or scenario.
- ☐ Use algebraic properties to explain the steps in an expression or equation.

Lesson 2: Part B
- ☐ Solve and graph solutions for single-variable compound inequalities.
- ☐ Solve and graph solutions for single-variable inequalities that contain absolute value.
- ☐ Identify inequalities containing absolute value as having no solution or as true for all real numbers.

Lesson 3: Part A
- ☐ Solve a multi-step equation.
- ☐ Use substitution to prove your solutions are correct.
- ☐ When given a problem with defined variables, write a single-variable equation.
- ☐ When given an equation that has more than one variable, solve for a specific variable.

Lesson 3: Part B
- ☐ Determine equivalent ratios.
- ☐ Solve a proportion using cross-products.

| UNIT 1 | Record Keeping | Name: | | | | | | Algebra 1 |

Lesson		Part		Guided Notes	Practice 1	Practice 2	Mastery Check	Lesson Test	Targeted Review
4	Solving Inequalities	A	Single-Variable Inequalities						
		B	Compound Inequalities						
5	Ratios, Proportions, and Rates	A	Ratios and Proportions						
		B	Unit Conversions						
6	Understanding Data	A	Data Calculations						
		B	Interpreting Data						

Unit Test I   Date _____   Score _____

Lesson Objectives

Lesson 4: Part A
- ☐ Identify numbers as real, rational, integers, whole, natural, and/or irrational.
- ☐ Draw a scaled number line showing numbers and approximate numbers.
- ☐ Explain why
  - a rational number + a rational number = a rational number
  - a rational number + an irrational number = an irrational number
  - an irrational number · an irrational number = an irrational number

Lesson 4: Part B
- ☐ Determine when an absolute value equation has no solution or is an identity.

Lesson 5: Part A
- ☐ Identify algebraic properties within an equation or scenario.
- ☐ Use algebraic properties to explain the steps in an expression or equation.

Lesson 5: Part B
- ☐ Solve and graph solutions for single-variable compound inequalities.
- ☐ Solve and graph solutions for single-variable inequalities that contain absolute value.
- ☐ Identify inequalities containing absolute value as having no solution or as true for all real numbers.

Lesson 6: Part A
- ☐ Rewrite equations with rational coefficients to integer coefficients before solving.
- ☐ Determine that an equation has no solution or is an identity.

Lesson 6: Part B
- ☐ Convert units for a value using a single conversion.
- ☐ Convert units for a value using multiple conversions.
- ☐ Convert compound units for a value.

Lesson 1
The Language of Algebra

Outline

Part A
Real Numbers

- Rational Numbers and Irrational Numbers
- Classifying Real Numbers

Part B
Algebraic Properties

- Commutative and Associative Properties
- Inverse, Identity, and Zero-Product Property
- Distributive Property
- Properties of Equality

Targeted Review

Vocabulary

- real numbers ($\mathcal{R}$)
- rational numbers ($\mathcal{Q}$)
- integers ($\mathcal{Z}$)
- whole numbers ($\mathcal{W}$)
- natural numbers ($\mathcal{N}$)
- irrational numbers ($\mathcal{I}$)
- term
- like terms

Part A: Real Numbers

Objectives

In this part of the lesson, you will learn about real numbers.

By the end of this lesson you will be able to do the following:

- ✓ Identify numbers as:
  - real
  - integers
  - whole
  - rational
  - irrational
  - natural

- ✓ Draw a scaled number line showing numbers and approximate numbers.

- ✓ Explain why:

  a rational number + a rational number = a rational number.

  a rational number + an irrational number = an irrational number.

  an irrational number · a rational number = an irrational number.

Why?

It is important to understand the language of algebra. If the directions tell you that all solutions are integer values or all solutions are irrational, you need to be familiar with those words so that you can determine if you have the correct answer.

Warm Up

Fill in the word from the list that matches the definition.

algebra	coefficient	integer	terminating	variable

_____ 1) a branch of mathematics that deals with numbers that may be represented by variables

_____ 2) a letter that represents an unknown quantity or number

_____ 3) a quantity (often a number) placed in front of a variable in an expression

_____ 4) positive and negative whole numbers, $\{..., -3, -2, -1, 0, 1, 2, ...\}$

_____ 5) ending

How many of these words did you already know? Whether new or a review, you will use these words throughout your exploration of algebra.

Now that your mind is ready for math, it is time to explore something new. In this part of the lesson, you will watch videos, fill in guided notes, and confirm your understanding at each checkpoint.

EXPLORE 1A

▶ Rational and Irrational Numbers

- All numbers in Algebra 1 are _____ numbers (ℝ).

- Each set of numbers has a letter that can be used as _____ for that set.

- The _____ diagram is a visual way to categorize number sets because a number can *only* be rational or irrational, never both.

- The first two subsets of _____ are _____ and _____.

The Real Number System

Q	I

Rational numbers

- Rational numbers can be written as a _____ of integers.

- When you add, subtract, multiply, or divide a _____ the result is a rational number. Rational numbers are _____ under the four operations.

- Using symbols, this means that:

  $Q + Q = Q$ $Q - Q = Q$ $Q \cdot Q = Q$ $Q \div Q = Q$

- In other words, if you _____ with a rational number, your _____ will be a rational number.

- Rational numbers can be written as fractions or decimals. However, when rational numbers are written as decimals they will terminate, or be a repeating decimal.

  A) $\frac{1}{4} = 0.25$　　　　　**B)** $\frac{1}{9} = 0.1111... = 0.\overline{1}$

1A EXPLORE

Irrational Numbers

- With irrational numbers, there is _____ or _____ that is exactly equal to the decimal value.

- This is because irrational numbers are _____ and _____ when written in decimal form.

- The sum or product of a rational and irrational number will *always* be irrational *except* when multiplying by _____.

- Assuming all values are non-zero, these expressions are *always* true:

    -
    -
    -

- Assuming all values are non-zero, this expression may only be true *sometimes*: _____

Example 1

Determine whether the expression will be rational or irrational using the mathematical shorthand {Q, I}. Then place each number in the diagram in your Guided Notes.

A) $5 + 0.\overline{3}$

$Q + Q = Q$

B) $-3 - 0.75077507775...$

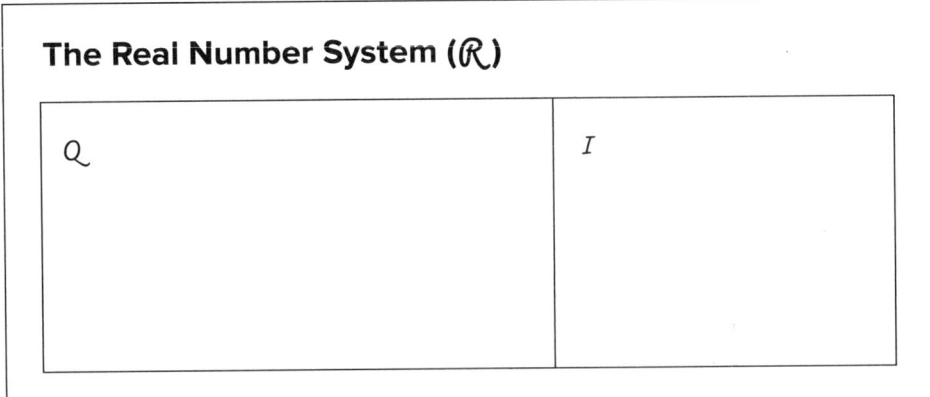

The Real Number System ($\mathcal{R}$)

| Q | I |

EXPLORE 1A

Example 2

Place the numbers on the Real Number diagram. Then order them on a number line.

A) -50% B) $\frac{8}{8}$ C) π D) $-\frac{6}{3}$ E) 0.75

It may be helpful to write the numbers in the same form, in this case all decimals.

The values range from -2 to approximately 3.14. The number line should extend slightly past these numbers to show a better approximation of the values. You must label the points on the number line with the given numbers; not the decimal values.

> In this course when writing answers as a fraction, mixed fractions will be used less frequently. For example, if your answer is $\frac{6}{4}$, you should write $\frac{3}{2}$ (rather than $1\frac{1}{2}$).

☑ Checkpoint

Order the given numbers on the number line. Next to the number label with Q for rational or I for irrational.

A) $-\frac{3}{2}$ B) $\sqrt{5}$ C) -150% D) $\frac{3}{4}$

1A EXPLORE

▷ Classifying Real Numbers

- As you know, Real Numbers have two main subsets: _____ and _____ _____.

- Rational numbers are also made up of a subset of numbers.

    - Integers _____: positive and negative whole numbers, or _____

    - Whole numbers _____: set of numbers that begin with zero, or _____

    - Natural numbers _____: set of numbers that begin with one, or _____

Example 3

Use the real number diagram to classify the given number by all of the sets to which it belongs using mathematical shorthand.

A) -5 $\{Z, Q, R\}$

B) $\frac{2}{3}$ _____

C) 0 _____

D) $\sqrt{4}$ _____

E) $\sqrt{2}$ _____

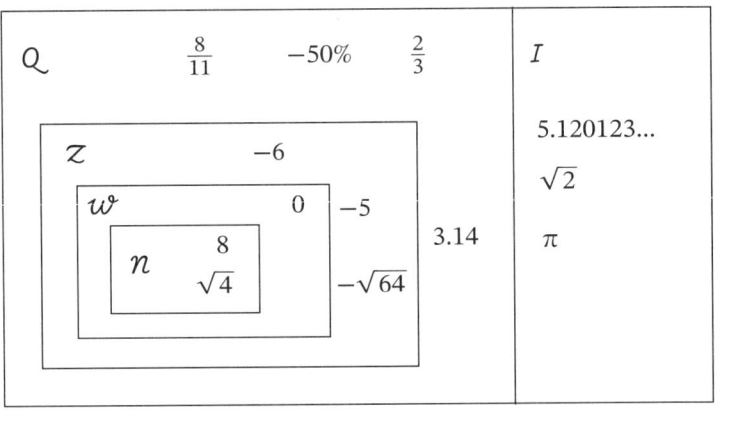

The Real Number System (R)

Q, $\frac{8}{11}$, -50%, $\frac{2}{3}$, I
Z, -6, 5.120123...
W, 0, -5, $\sqrt{2}$
N, $\frac{8}{\sqrt{4}}$, $-\sqrt{64}$, 3.14, π

Why is it important to learn math vocabulary in Algebra 1?

Example 4

Label each section of the real number diagram with the correct classification. Then place the numbers for A–H onto the diagram correctly.

A) $\sqrt{7}$ **B)** 0 **C)** 100% **D)** $\frac{10}{5}$ **E)** -16 **F)** $\frac{22}{7}$ **G)** 6.3 **H)** π

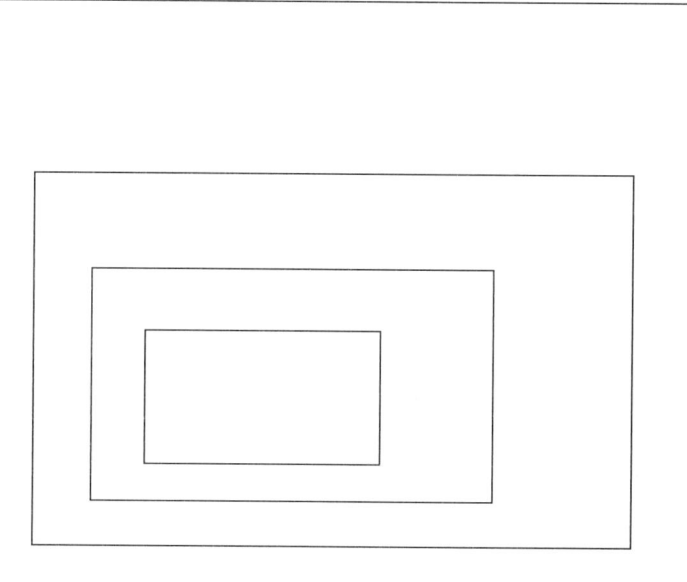

☑ Checkpoint

State the most specific classification for the given numbers. Write out the word and the mathematical shorthand.

A) 12

B) $\sqrt{12}$

C) -12

1A PRACTICE 1

✏️ Practice 1

Complete the problems on a separate sheet of paper.

Classify each number by all sets to which it belongs using math shorthand.

1) $\sqrt{81}$

2) 0

3) $\frac{8}{11}$

4) $5.12012301234012345...$

5) $-\sqrt{36}$

6) Draw the real number diagram and label with the correct classifications, then place the numbers from problems 1–5 on the diagram.

Classify each number by the most specific set to which it belongs. You will graph these points in problem 11.

7) Point A: $-24\left(\frac{1}{12}\right)$

8) Point B: $\pi - 3$

9) Point C: $\frac{2}{3} - 1$

10) Point D: $\sqrt{25} - \sqrt{16}$

11) Draw a number line to graph problems 7–10 as points A, B, C, and D.

Finish the sentence with one of the following words: always, sometimes, never.

12) (irrational) + (irrational) is _____ irrational.

13) (rational) · (rational) is _____ irrational.

14) (rational) + (irrational) is _____ irrational.

MASTERY CHECK 1A

Mastery Check

Show What You Know

A) Place the given values on the number line.

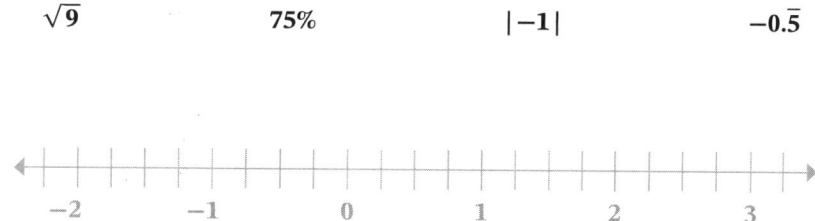

$\sqrt{9}$ 75% $|-1|$ $-0.\overline{5}$

B) Grant and Jacob are classifying the given numbers. Grant says that $-0.\overline{5}$ is rational. Jacob says that $-0.\overline{5}$ is irrational because it is a non-terminating decimal. Who is correct? Explain.

C) Grant and Jacob decide they want to find the sum of all the given values. Without finding the sum, explain if the solution will be rational or irrational.

D) Complete the sentence with the word always, sometimes, or never. Explain.

(any irrational number) + (any rational number) is _____ rational.

Say What You Know

In your own words, talk about what you have learned using the objectives for this part of the lesson and your work on this page.

Practice 2

Complete the problems on a separate sheet of paper.

Classify each number by all sets to which it belongs using math shorthand.

1) $\sqrt{2}$

2) $\frac{39}{3}$

3) $-|11|$

4) 75%

5) $2{,}352$

6) Draw the real number diagram and label with the correct classifications, then place the numbers from problems 1–5 on the diagram.

Classify each number by the most specific set to which it belongs. You will graph these points in problem 11.

7) Point P: $\frac{5}{2} + \frac{1}{2}$

8) Point Q: $\frac{1}{9} - \frac{5}{9}$

9) Point R: $\sqrt{5}\,(0)$

10) Point S: $\pi - \pi$

11) Draw a number line to graph problems 7–10.

12) Name all of the sets that fall under rational numbers.

Part B: Algebraic Properties

Objectives

In this part of the lesson, you will learn about algebraic properties.

By the end of this lesson, you will be able to do the following:

- Identify algebraic properties within an equation or scenario.
- Use algebraic properties to explain the steps in an expression or equation.

Why?

Being able to use the correct algebraic properties from your Formula Sheet to solve problems is foundational to algebra. Being able to explain why you are using them completes the problem-solving plan and deepens your understanding of algebra.

Warm Up

Find the formula on your Formula Sheet.

1) Write the formula for the Surface Area of a rectangular prism.

2) How many feet are in one mile?

3) What is the formula for Standard Form of a Linear Equation?

Using the correct formula is the basis for solving algebraic problems. Using your Formula Sheet can help you make sure you are starting on the right path.

▷ Commutative and Associative Properties

- With _____ you can simplify expressions to solve equations.

- A _____ is an accepted rule of mathematics when working with terms.

- A single number, variable, or the product of a number and a variable is a _____.

- The _____ Property can be demonstrated by $a + b = b + a$
$$ab = ba$$

- The _____ Property can be demonstrated by $a + (b + c) = (a + b) + c$
$$a(bc) = (ab)c$$

- The properties work for all real numbers because _____ are used in Algebra 1.

1B EXPLORE

- The Commutative Property is true for the math operation(s) of: _____.

- The Associative Property is true for the math operation(s) of: _____.

Example 1
Determine the property being demonstrated.

A) Amelie was given the expression $8 + 9 + 2$. She wrote $8 + 2 + 9$ and still found the correct sum.

   Commutative Property

B) Ivan was given the expression $(3 \cdot 7)(5)$. He decided to simplify $(3)(7 \cdot 5)$.

C) Chip was given the expression $-6 + 4 + (-2)$. He wrote $-2 + (-6) + 4$ and still found the correct answer.

☑ Checkpoint

Using variables to represent real numbers, create your own example equations demonstrating the Commutative and Associative Properties for Multiplication.

▶ Inverse, Identity, and Zero-Product Property

- _____ Property $a + 0 = a,$
                                     $a \cdot 1 = a$

- _____ Property $a \cdot 0 = 0,$
                                     If $ab = 0$, then either a or b equals zero.

  - The Zero-Product Property is true for the math operation(s): _____.

- _____ Property $\dfrac{a}{b} \cdot \dfrac{b}{a} = 1,$
                                     $a + (-a) = 0$

  - The _____ Property is used when adding a number and its opposite together to "undo" a given term.

  - A number and its reciprocal that multiply to one is called a _____.

EXPLORE 1B

Example 2

Name the property being used and whether addition or multiplication is being used.

A) $3 \cdot \frac{1}{3} = 1$

B) $26 + 0 = 26$

☑ Checkpoint

Name the property being demonstrated. Answers can be used more than once.

Inverse Property   Identity Property   Zero-Product Property

A) $x + (5 + -5) = x$ **B)** $3b \cdot (0) = 0$

C) $\left(\frac{8}{8}\right)m = m$ **D)** $\left(\frac{1}{2}\right)(2) = 1$

E) $11n - 11n = 0$

▶ Distributive Property

- The Distributive Property _____ one term by a set of terms found inside _____ symbols.

- The Distributive Property is the only property that uses a _____ symbol.

- The Distributive Property states that: _____
  _____

- _____ are terms with the same variable raised to the same power.

Example 3

Simplify the expression using the Distributive Property.

A) $7(2x - 11)$

$7(2x) - 7(11)$ ◀ Each term in the parentheses will be multiplied by 7.

$14x - 77$ ◀ This is the simplified expression because there are no more like terms.

B) $x(y + 8)$

> When there is more than one variable, write the variables alphabetically. When a variable and number are multiplied together, write the number then the variable.

1B EXPLORE

✓ Checkpoint

Name the property being described.

A) $4(a + 3) = 4a + 12$

B) $-15 + 15 = 0$

C) $82(0) = 0$

▶ Properties of Equality

- The Reflexive Property states that: _____.

- The Property of Symmetry states that: _____.

- The _____ Property states that:

  If $a = b$, then b can replace a in expressions and equations.

- The _____ says: If $a = b$, then $a + c = b + c$.

- When using the Addition Property of Equality, the same number is _____ to both sides to maintain _____.

- The _____ says: If $a = b$, then $ac = bc$.

- When using the Multiplication Property of Equality, _____ of the equation must be multiplied by the _____ to maintain equality.

Example 4

Name the property that justifies each step.

$\frac{2}{9}x - 6 = 1$ ◂ Given

$\frac{2}{9}x - 6 = 1$
$\phantom{\frac{2}{9}x}+6 +6$ ◂ Addition Property of Equality (add 6 to both sides of the equation)

$\frac{2}{9}x = 7$ ◂ Inverse Property for Addition $(-6 + 6)$

$\left(\frac{9}{2}\right)\frac{2}{9}x = 7\left(\frac{9}{2}\right)$ ◂ Multiplication Property of Equality (multiply both sides of the equation by the same value)

$x = \frac{63}{2}$ ◂ Inverse Property for Multiplication $\left(\text{simplify } \left(\frac{9}{2}\right)\left(\frac{2}{9}\right) \text{ to find the value of } x\right)$

Example 5

Name the property that justifies each step.

$7 - 2c = -3(c + 1)$ ◂ Given

$7 - 2c = -3c - 3$ ◂

$ + 2c + 2c$

$ 7 = -c - 3$ ◂

$ 7 = -c - 3$
$+ 3 + 3$ ◂

$ 10 = -c$ ◂

$10(-1) = -1(-c)$ ◂

$ c = -10$ ◂

☑ Checkpoint

Name the property being described. Use your Formula Sheet.

A) Caleb says that $4 = m$ is the same as $m = 4$.

B) Jimena was given the equation $\frac{5}{4}x = 10$. Her first step was to multiply both sides by $\frac{4}{5}$.

C) Albie checked Jimena's work by substituting $x = 8$ into the equation $\frac{5}{4}x = 10$. His result was $10 = 10$. What *two* properties does this demonstrate?

Practice 1

Complete the problems on a separate sheet of paper.

Determine the property being used.

1) Milo was given the expression $9 \cdot 5$. Milo wrote $5 \cdot 9$.

2) Gemma needs to determine the reciprocal of $\frac{7}{8}$. What property will she use?

3) Kevin knows that any number multiplied by zero is zero. What property is this?

4) Calen simplified the expression $5(x - 12)$ to $5x - 60$. What property is this?

5) Maria was asked to write an example of a number added to its opposite is zero. She wrote: $-3 + 3 = 0$.

6) After working on the first step of an equation, Sharla had $6x + 0 = 5$, then wrote $6x = 5$.

7) Sam was solving the problem $3x + 5 = 9$. They added -5 to both sides of the equation.

8) Given the problem $3x = 4$, Qorban wanted to multiply each side by $\frac{1}{3}$.

9) Joseph solved an equation and found that $x = 12$. He replaced x with 12 to check his answer.

10) Mille wrote $n = 16$, Connier wrote $16 = n$. Which property says they have the same answer?

---

11) Name the property to justify each step. Use your Formula Sheet.

$$5x - 6 = 14 \quad \blacktriangleleft \text{ Given}$$

A)
$$5x - 6 = 14$$
$$+6 \quad +6$$
$$5x = 20$$

B) $\left(\frac{1}{5}\right)(5x) = \left(\frac{1}{5}\right)(20)$
$$x = 4$$

12) Name the property to justify each step.

$$11 = \tfrac{1}{3}x + 8 \quad \blacktriangleleft \text{ Given}$$

A)
$$11 = \tfrac{1}{3}x + 8$$
$$-8 \quad\quad -8$$
$$3 = \tfrac{1}{3}x$$

B) $(3)(3) = (3)\left(\tfrac{1}{3}x\right)$
$$9 = x$$

C) $x = 9$

Mastery Check

✎ Show What You Know

A) Finish each sentence with one of the following words: always, sometimes, never.

- All integers are _____ rational numbers.

- ℝ numbers are _____ integers.

B) Finish each sentence with one of the following words: always, sometimes, never.

- The associative property is _____ true for division.

- The Identity Property for Addition _____ equals zero.

- The Distributive Property _____ uses subtraction.

- The Multiplication Property of Equality _____ multiplies both sides of an equation by the same number.

ᴴᴵ Say What You Know

In your own words, talk about what you have learned using the objectives for this part of the lesson and your work on this page.

1B PRACTICE 2

✏ Practice 2

Complete the problems on a separate sheet of paper.

1) DeAnne was asked to write an example for a number multiplied by its reciprocal is one. She wrote: $-\frac{5}{6}\left(-\frac{6}{5}\right) = 1$

2) Hunter wanted to quickly add the expression $(9 + 8) + 2$. He decided to write $9 + (8 + 2)$. What property is demonstrated?

3) Elle was given the expression $-3 + 6 + 3$. She wrote $-3 + 3 + 6$. What property is this?

4) Elle decided to simplify the expression $-3 + 6 + 3$ another way. She wrote $0 + 6$ as her first step. What property did she use?

5) Fahrid said that $4 \cdot \frac{8}{8} = 4$. What property does this show?

6) The directions on Zeke's math lesson said to simplify. He was given $x(y + 8)$. What property can he use to simplify this?

7) What property allows you to multiply an equation by the same number on both sides?

8) If you solve an equation and want to check that your answer is correct, what property would you use to check your answer?

9) Betsy solved an equation and found the solution $-4 = m$. Patty found the solution to be $m = -4$. What property shows this is the same solution?

10) The equation $34x + 95 = 5$ was given. Gary's first step was to add -95 to both sides. What property did Gary use?

---

11) Name the property to justify each step.

$$4(x + 7) = -3 \qquad \blacktriangleleft \text{Given}$$

A) $\quad 4x + 28 = -3$

B) $\quad \begin{array}{r} 4x + 28 = -3 \\ -28 \quad -28 \end{array}$

$\quad 4x = -31$

C) $\left(\frac{1}{4}\right)(4x) = \left(\frac{1}{4}\right)(-31)$

$\quad x = -\frac{31}{4}$

12) Name the property to justify each step.

$$-\frac{1}{6}x = 12 \qquad \blacktriangleleft \text{Given}$$

A) $(-6)\left(-\frac{1}{6}x\right) = (-6)(12)$

$\quad x = -72$

Check:

B) $\quad -\frac{1}{6}(-72) = 12$

C) $\quad 12 = 12$

Targeted Review

In the Targeted Review, you will practice topics you have mastered in earlier lessons. Reviewing these concepts will help you be successful as you work through this unit.

Complete the problems on a separate sheet of paper.

Evaluate.

1) $|-6| - |4|$

2) $|-6 - 4|$

Simplify. Name the least common denominator.

3) $\frac{4}{5} + \frac{2}{3}$

4) $\frac{1}{8} + \frac{2}{5} - \frac{1}{10}$

5) $-\frac{3}{4} + \frac{2}{3}$

6) $\frac{12}{5} - \frac{2}{3} - \frac{1}{2}$

Determine the value of the expression.

7) $\sqrt{81}$

8) $\sqrt{25}$

Simplify using order of operations.

9) $|-8| + 2(6)\left(\frac{1}{4}\right) - \left(\sqrt{25} \div \frac{1}{5}\right)$

10) $9\left(\frac{2}{3}\right) - |5 - 11| + (-3)(13)$

Solve.

11) $\frac{1}{5}x = -8$

12) $12 - x = -3$

13) $\frac{x}{2} + 6 = -5$

14) $4x - 3 = 7$

Multiple Choice

15) What is the value of the expression, $2c^2 - c$, when $c = -3$?

   A) -15
   B) -9
   C) 15
   D) 21

16) Select *all* expressions that are equivalent to $5x$.

   ☐ $x + 4x$
   ☐ $x + 5$
   ☐ $2x + 3x$
   ☐ $10x - 5x$

Lesson 2
Solving Equations

Outline

Part A
Multi-Step Equations

- Solving Multi-Step Equations
- Variables on Both Sides of the Equation
- Defining Variables in Word Problems
- Solving Equations with More than One Variable

Part B
Equations with Special Cases

- Rewriting Equations with Integer Coefficients
- One Solution, No Solution, or All Real Numbers

Targeted Review

Vocabulary

- no solution
- identity

Part A: Multi-Step Equations

Objectives

In this part of the lesson, you will learn about multi-step equations.

By the end of this lesson, you will be able to do the following:

- ✓ Solve a multi-step equation.
- ✓ Use substitution to prove your solutions are correct.
- ✓ Write a single-variable equation when given a problem with defined variables.
- ✓ Solve for a specific variable when given an equation that has more than one variable.

Why?

Solving for a variable in multi-step equations is a foundation for algebra. Mastering multi-step equations is an important skill for success in secondary math.

☁ Warm Up

1) Label the equation with the words coefficient, expression, and variable.

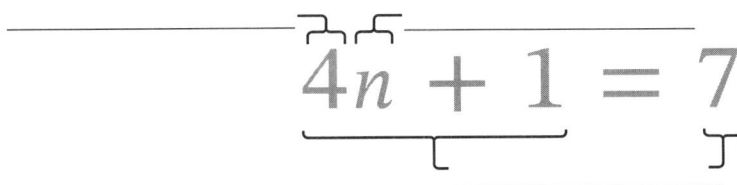

2) Name the inverse operation.

   A) The inverse of division is _____.

   B) The inverse of addition is _____.

Did you remember all the parts of an equation? What about inverse operations?

Having a solid foundation of basic algebraic language will help as you build on that knowledge.

▶ Solving Multi-Step Equations

- When you solve problems in Algebra 1, _____ is how you will approach the problem.

- During the Implement stage of the problem solving method in Algebra 1, you will _____ your plan and _____ your work.

- When solving a problem, you should be able to _____ why your answer makes sense.

EXPLORE 2A

- _____ are used to solve equations by rearranging terms on both sides of the equals sign (=).

- When solving for a variable in an equation, the goal is to _____ the variable.

Example 1

Use the Plan, Implement, Explain method to solve the equation. Check your solution.

$-5x + 8 = 12$

Plan $\cdot(-5)$ ↑ Move backwards through the plan,
$+(8)$ using inverse operations.

Implement

$-5x + 8 = 12$
$ -8 -8$
$-5x = 4$
$\left(-\frac{1}{5}\right)(-5x) = \left(-\frac{1}{5}\right)(4)$
$x = -\frac{4}{5}$

Explain

◂ Addition Property of Equality

◂ Multiplication Property of Equality

Check

$-5x + 8 = 12$
$-5\left(-\frac{4}{5}\right) + 8 = 12$
$4 + 8 = 12$
$12 = 12$ ✓

Example 2

Solve. The Plan portion is completed for you.

$\frac{1}{3}(2x + 7) = -4$

Plan $\cdot(2)$ ↑ Work backwards through
$+(7)$ the steps to isolate x.
$\cdot\left(\frac{1}{3}\right)$

Implement   **Explain**   **Check**

2A EXPLORE

Example 3

Solve.

$\frac{2}{5}(1-3q) = 6$

Plan $\cdot(-3)$
             $+(1)$
             $\cdot\left(\frac{2}{5}\right)$

Implement   **Explain**

☑ Checkpoint

Plan, Implement, and Explain to solve.

$\frac{4}{5}(2x+3) = 12$

Variables on Both Sides of the Equation

- As you become more confident using Plan, Implement, Explain to solve multi-step equations, you will likely no longer need to write out the _____ stage.

- Equations can be solved in more than one way, but the _____ will be the same.

- If a variable is found on both sides of an equation, first determine the side of the _____ on which you want to keep the variable.

Example 4

Solve.

$8(2x - 1) + 6 = 11x + 7 - 4$

Implement

$8(2x - 1) + 6 = 11x + 7 - 4$

$16x - 8 + 6 = 11x + 3$ ◀ Distribute and simplify terms

$16x - 2 = 11x + 3$

$-11x -11x$ ◀ Addition Property of Equality

$5x - 2 = 3$

$+2 +2$ ◀ Addition Property of Equality

$\left(\frac{1}{5}\right)(5x) = \left(\frac{1}{5}\right)(5)$ ◀ Multiplication Property of Equality

$x = 1$

Explain

Check

$8(2(1) - 1) + 6 = 11(1) + 7 - 4$

$8(2 - 1) + 6 = 11 + 7 - 4$

$8 + 6 = 18 - 4$

$14 = 14$ ✓

2A EXPLORE

Example 5

Solve.

$3x + 8 = 5x + 4$

Implement   **Explain**

☑ Checkpoint

Solve.

$11x - 2(x + 6) = 51$

▶ Defining Variables in Word Problems

- _____ allow formulas to be written to generalize the way that numbers relate to one another.

- When solving word problems that do not have a standard formula, the _____ who is solving the problem _____ them.

EXPLORE 2A

Example 6

Define a variable, then write and solve your equation.

To make sure a program was correct, the first scientist ran 48 initial experiments. After that, the first scientist ran three experiments every day. A second scientist ran 15 initial experiments of a program. After that, the second scientist ran four experiments per day. How many days will it take for both scientists to run the same number of experiments?

d: days experiments are running.

$$48 + 3d = 15 + 4d$$
$$-3d -3d$$
$$48 = 15 + d$$
$$-15 -15$$
$$33 = d$$

Check

$$48 + 3(33) = 15 + 4(33)$$
$$48 + 99 = 15 + 132$$
$$147 = 147 \checkmark$$

Explain

> For word problems, Explain is more about the meaning of the answer and less about justifying using the algebraic properties. By answering with a sentence that resolves the problem, you complete the Explain step.

Example 7

Write and solve an equation to find the three integers.

The sum of three consecutive integers is −57.
Write and solve an equation to find the three integers.

> In this problem, consecutive means three numbers in order, for example 1, 2, 3.

n: first number
$n + 1$: second number
$n + 2$: third number
$n + (n + 1) + (n + 2) = -57$

> In this course, answers are usually written from least to greatest.

2A EXPLORE

> ☑ **Checkpoint**
>
> Why is it important to define the variable or variables you use when writing an equation?

▶ Solving Equations with More than One Variable

- When equations have more than one variable, the goal shifts from solving for an exact value to _____ a specific variable.

- Unless specified, when solving equations that have more than one variable in this course, all variables are _____.

- The directions, "Solve in terms of x," means to isolate, or get x _____ on one side of the equation.

- Another way to represent isolating variables is $3x + 4 = 5y; x$

  The variable after the semicolon is what needs to be isolated.

Example 8

Solve the equation in terms of y.

$Ax + By = C$

Plan $\cdot (B)$
$+ Ax$

Implement	**Explain**
$Ax + By = C$	
$\underline{-Ax \qquad\qquad -Ax}$	◂ Addition Property of Equality
$By = C - Ax$	
$\left(\frac{1}{B}\right)(By) = \left(\frac{1}{B}\right)(C - Ax)$	◂ Multiplication Property of Equality
$y = \frac{C - Ax}{B}$	
$y = \frac{-Ax + C}{B}$	◂ Commutative Property
OR $\quad y = \frac{-Ax}{B} + \frac{C}{B}$	

EXPLORE 2A

Example 9

Solve the equation in terms of b.

$$\frac{m - a + b}{5} = 3c$$

Example 10

Solve.

$$\frac{1}{3}(2R - r) = x;\ R$$

☑ Checkpoint

Max wanted to use the formula for the area of a triangle to find the height when given the area and the base. Write the formula and solve in terms of h.

Practice 1

Complete the problems on a separate sheet of paper.

Solve. Show Plan, Implement, Explain.

1) $\frac{x}{4} - 8 = -3$

2) $\frac{1}{4}(3x - 5) = -1$

Solve.

3) $2(b - 4) + 5 = 9$

4) $4v = 2v - 7 + 5$

5) $-2b + 5 - b + 3 = -12$

6) $7(q + 2) = 9q + 5$

Define a variable and write an equation for the given scenario. Solve your equation.

7) The perimeter of the rectangle equals 26 inches.
What is the length of the rectangle?
What is the width of the rectangle?

8) Emily and Reggie were comparing how much money they had each made from mowing lawns that week. Emily charged $15 per lawn, and she made $5 in tips. Reggie charged $10 per lawn, and made $15 in tips. After everything was counted, they discovered they made the same amount of money and each mowed the same amount of lawns. How many lawns did they each mow?

9) Three consecutive integers have a sum of −78. Write and solve an equation to find the three integers.

Solve in terms of the named variable. Remember to write your Plan.

10) $qx - 4 = c; x$

11) $d = rt; r$

12) The formula for the surface area of a cylinder is shown below. Solve the formula in terms of h.

$$SA = 2\pi r^2 + 2\pi rh$$

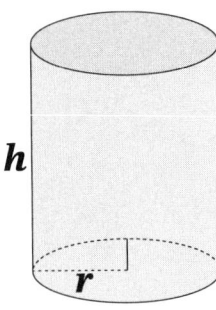

Mastery Check

✎ Show What You Know

A) Find the largest possible value for x by filling in the boxes using the natural numbers 1 through 9 and then solving for x.

$$1 \quad 2 \quad 3 \quad 4 \quad 5 \quad 6 \quad 7 \quad 8 \quad 9$$

- You can use each number only once in the equation.
- Repeat the process until you have the largest value for x.
- Show all of your work.
- (Hint: Do not erase. Rather, keep a log of all attempts.)

$$\frac{1}{\square}\left(x + \square\right) = \square$$

B) Explain why you believe your answer to x is the largest possible value.

ᷧ Say What You Know

In your own words, talk about what you have learned using the objectives for this part of the lesson and your work on this page.

2A EXPLORE

✎ Practice 2

Complete the problems on a separate sheet of paper.

Solve. Show Plan, Implement, Explain.

1) $\frac{2}{3}x + 15 = -1$
2) $\frac{5}{9}(x - 18) - 3 = 2$

Solve.

3) $5(2x + 3) - 8 = 65$
4) $\frac{3}{2}x - 11 = x - 4$
5) $17 = -12x + 3 - 1 - 3x$
6) $16(2x - 1) + 4 = -12$

Define a variable and write an equation for the given scenario. Solve your equation.

$(2x - 3)$ feet

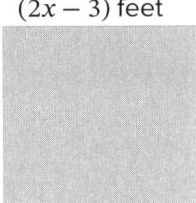

7) The perimeter of a square is 96 feet. Write an equation using the square to find the value of x and the length of each side. Write your answers as decimal values.

8) Four-fifths of a number less eight is equal to three-fifths of the same number plus fourteen. Write an equation and solve for the number.

9) Three consecutive even integers have a sum of 312. Write and solve an equation to determine the three integers.

Solve in terms of the named variable.

10) $10a - b = c; a$
11) $V = lwh; w$

12) The formula for the surface area of a square pyramid is given below. Write the formula in terms of l.

$$SA = s^2 + 2sl$$

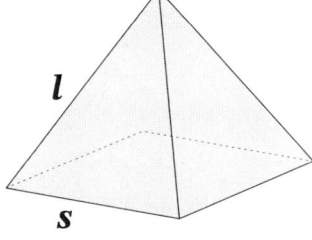

EXPLORE 2B

Part B: Equations with Special Cases

Objectives

In this part of the lesson, you will learn about equations with special cases.

By the end of this lesson, you will be able to do the following:

- Rewrite equations with rational coefficients to integer coefficients before solving.
- Determine that an equation has no solution or is an identity.

Why?

Understanding the different types of solutions to algebraic equations is important to determining if your solutions make sense.

☁ Warm Up

The Least Common Denominator (LCD) is the smallest multiple that fractions have in common.

Name the LCD, then evaluate.

1) $\frac{1}{6} - \frac{3}{4}$

2) $\frac{5}{7} + \frac{3}{5} - \frac{1}{2}$

Being able to find the least common denominator is important for solving some algebraic problems.

▶ Rewriting Equations with Integer Coefficients

- It is very helpful to know how to clear fractions and decimals from equations so that only _____ values remain.

- To clear fractions, you should find the _____.

- To clear decimals, you should _____ by the greatest place value of the decimals.

2B EXPLORE

Example 1

Solve.

$\frac{2}{3}x + \frac{1}{2} = \frac{3}{4}$

Plan   Clear fractions using LCD

Implement   **Explain**

$\frac{2}{3}x + \frac{1}{2} = \frac{3}{4}$

$12\left(\frac{2}{3}x + \frac{1}{2}\right) = 12\left(\frac{3}{4}\right)$ ◀ Multiplication Property of Equality

$8x + 6 = 9$ ◀ Distributive Property

$-6 \quad -6$ ◀ Addition Property of Equality

$8x = 3$

$\left(\frac{1}{8}\right)8x = \left(\frac{1}{8}\right)(3)$ ◀ Multiplication Property of Equality

$x = \frac{3}{8}$

Check

$\frac{2}{3}\left(\frac{3}{8}\right) + \frac{1}{2} = \frac{3}{4}$

$\frac{1}{4} + \frac{1}{2} = \frac{3}{4}$

$\frac{3}{4} = \frac{3}{4}$ ✓

Example 2

Solve. The first part of this example is completed for you.

$0.15n + 0.8 = 2$

Plan   Clear the decimals from this equation by multiplying by the greatest place value of the decimals.

Implement   **Explain**

$0.15n + 0.8 = 2$

$100(0.15n + 0.8) = 100(2)$ ◀ Multiplication Property of Equality / Distributive Property

EXPLORE 2B

✓ Checkpoint

Solve the equation by writing all numbers as integers. Show Implement.

Use the Distributive Property to solve for x.

$$\frac{2}{3}x + \frac{2}{5} = \frac{8}{15}$$

▶ One Solution, No Solutions, or All Real Numbers

- You can use substitution to determine if your solution to an equation with _____ is correct.

- For some equations there is no value, or _____ that makes the equation true.

Example 3

Solve.

$3x + 2 = 3(x - 5)$

Implement and Explain

$3x + 2 =$	$3(x - 5)$	
$3x + 2 =$	$3x - 15$	◂ Distributive Property
$-3x$	$-3x$	◂ Addition Property of Equality / Inverse
$2 =$	-15	

no solution

Check

You can replace the variable with any _____ and the equation will have a false equality statement.

2B EXPLORE

- Another possible solution to an equation is that the equation is true for _____.

- An equation with a solution that is all real numbers is also called an _____.

Example 4

Solve.

$18 = 6(x + 3) - 6x$

Implement

$18 = 6(x + 3) - 6x$

Check Use any _____ for the value of x and the sides of the equation will remain _____.

Example 5

Solve.

$2x - 16 = 2(x - 8)$

$$
\begin{aligned}
2x - 16 &= 2x - 16 \\
-2x & -2x \\
+16 & +16 \\
0 &= 0
\end{aligned}
$$

All terms on both sides are exactly the same. All real numbers, ℝ can be written at any step.

☑ Checkpoint

Determine if x has no solution, one solution, or is true for all real numbers.

$-4x - 4 = -4(x + 1)$

38 Lesson 2: Solving Equations > Part B: Equations with Special Cases > Explore   Algebra 1 Student Worktext

Practice 1

Complete the problems on a separate sheet of paper.

Solve.

1) $2g - \frac{2}{5} = \frac{7}{5}$

2) $0.5h + 0.15 = 0.43h - 0.34$

3) $\frac{5}{8}x + 1 = \frac{1}{3}(x - 2)$

4) $0.4z + 0.24 = 1.5 + 0.1z$

5) Two-thirds of a number (n) plus three is the same as one-fourth of the same number minus one. Write and solve your equation.

6) Barbara spent $15.51 at a local farmers' market. She spent $5.26 on tomatoes and then bought p pounds of fruit for $4.10 per pound. How many pounds of fruit did Barbara buy?

7) Explain what an equation with no solutions will look like when solved.

Solve. Determine if the equation has one solution, no solutions, or has a solution of all real numbers.

8) $49 = 1 - (x + 1)$

9) $5(2x + 3) = 8x - (-2x + 6)$

10) $\frac{1}{4}(12x - 8) = 3x - 2$

11) $0.70k + 0.71 = 0.70k + 0.68$

12) Write your own example of an equation with a solution of all real numbers.

2B MASTERY CHECK

✅ Mastery Check

✎ Show What You Know

A) Solve the equation by first writing numbers as integers.
Determine if the equation has one solution, no solutions, or is an identity.

$$\frac{3}{10}(x - 20) = \frac{1}{2}x - \frac{1}{5}x - 6$$

B) Rian says that *all rational numbers* will make this equation true. Jo says that *all irrational numbers* will make this equation true. Explain how both students are correct.

ᵢlᵢlᵢ Say What You Know

In your own words, talk about what you have learned using the objectives for this part of the lesson and your work on this page.

Practice 2

Complete the problems on a separate sheet of paper.

Solve the equation by first writing the coefficients as integer values.

1) $\frac{2}{3}x - 3 = \frac{7}{3}x + 8$

2) $0.1h - 0.2 = 1.4$

3) $-\frac{2}{3}x + \frac{5}{4} = \frac{1}{2}$

4) $5.75 - 1.50 - 0.17p = 0$

5) Five-fourths of a number, less twenty-one divided by two is equal to one-fifth of the same number.

6) Mrs. Trainer and Mr. Rodriguez took a poll comparing food preferences among students in their classes. In Mrs. Trainer's class, 2 more than $\frac{3}{8}$ of the students prefer sweet potatoes over peas. In Mr. Rodriguez's class, 5 more than $\frac{1}{4}$ of the students preferred sweet potatoes. The two teachers have the same amount of students in the class and the same number of students who prefer sweet potatoes. How many students are in each class?

7) Explain what an equation that is an identity will look like when solved.

Solve. Determine if the equation has one solution, no solutions, or is an identity.

8) $2(b - 4) = 2b - 8$

9) $3p + 5 = 5p - 2p - 2$

10) $2n + 7 = 3n - 2$

11) $\frac{6}{7} - \frac{2}{3}v + \frac{1}{7} = \frac{2}{3}v + \frac{3}{7} - 1\frac{1}{3}v + \frac{4}{7}$

12) Write your own equation with no solution.

TARGETED REVIEW 2

Targeted Review

In the Targeted Review, you will practice topics you have mastered in earlier lessons. Reviewing these concepts will help you be successful as you work through this unit.

Complete the problems on a separate sheet of paper.

Classify the number using ALL of the sets to which it belongs using math shorthand.

1) -3.7

2) -27

Determine if the expression will result in a rational or irrational number.

3) $\sqrt{25} + \frac{1}{\sqrt{9}}$

4) $\sqrt{3} + 6$

Name the property being demonstrated.

5) $x + (-5 + 5) = x + 0$

6) $4n + 8 = 7 + 8$

Name the *two* properties being demonstrated.

7) $(0)(8x + 7y - z) = 0$

8) $9 + (3 + 1) = (9 + 1) + 3$

Draw a number line to order the given numbers.

9) P: 0   Q: $\frac{2}{3}$ R: $-\frac{1}{4}$ S: $-\frac{\pi}{\pi}$

Simplify the expression when $m = 6, p = -4$.

10) $m + (p - m) - p$

11) $-(mp) + m$

Draw and graph the ordered pairs on a coordinate plane.

12) A $(2, -3)$ B $(-3, 2)$ C $(4, 0)$ D $(0, 4)$ O $(0, 0)$

13) The perimeter of a rectangular shaped garden was 18 feet. The length of the garden was 6 feet. What is the width? Use your Formula Sheet to write the necessary formula, define the variables, and then solve.

14) The volume of a rectangular box was found to be 24 cubic feet. The height and the width of the box were 2 feet. What is the length of the box? Use your Formula Sheet to write the necessary formula, define the variables, and then solve.

Multiple Choice

15) Select all expressions that will result in a *rational* answer by filling in the appropriate box(es).

☐ $\sqrt{4} + \sqrt{5}$

☐ $\frac{1}{2} + \frac{11}{13}$

☐ $\pi \cdot \pi$

☐ $\frac{8}{\pi} \cdot \pi$

_____ 16) Samson reads 90 pages in his book each day. Kelley reads 20% more pages than Samson every day. How many pages does Kelley read daily?

A) 18 pages

B) 20 pages

C) 108 pages

D) 110 pages

Lesson 3
Solving Absolute Value Equations

Outline

Part A
Absolute Value Equations

- One- and Two-Step Absolute Value Equations
- Graphing Solutions Using the Midpoint
- Writing Absolute Value Equations

Part B
Multi-Step Absolute Value Equations

- Multi-step Absolute Value Equations
- Absolute Value Equations with No Solution and All Real Numbers

Targeted Review

Vocabulary

- absolute value

3A EXPLORE

Part A: Absolute Value Equations

Objectives

In this part of the lesson, you will learn about absolute value equations.

By the end of this lesson, you will be able to do the following:

- Solve one- and two-step absolute value equations.
- Graph one- and two-step absolute value equation solutions on a number line.
- Explain why an absolute value equation can have two solutions.
- Write and solve an absolute value equation from a given problem.

Why?

Solving and graphing absolute value equations shows that distance cannot be negative; however, the two solutions are equal distance from a midpoint on a number line in either direction.

Warm Up

Use the figure to answer problems 1–3.

1) Ed walked to Dee's house. How far did he travel?

2) Ed walked from Dee's house to Nancy's house. How far did he travel?

3) If Ed's house was at zero on the number line where is Dee's house? Nancy's house?

One- and Two-Step Absolute Value Equations

- The magnitude of a number is called _____.

- Absolute value (| |) can be thought of as _____ without direction.

    $|4|$ = _____ $|-4|$ = _____

EXPLORE 3A

- The absolute value of a single term is the distance between that term and _____.

- Since opposite terms have the same absolute value, the absolute value of an unknown has _____ possible solutions.

- When graphing absolute values on a number line, you must move both directions from the _____ to find both possible solutions.

- There are two _____ to consider when solving absolute value equations algebraically.

Example 1

Solve.
$|x - 5| = 7$

Plan Solve both cases of the absolute value equation.

Implement

$|x - 5| = 7$ ◁ **Explain** ◁ Given

Case 1:		Case 2:	
$x - 5 = 7$	**OR**	$-(x - 5) = 7$	◁ Definition of Absolute Value
		$x - 5 = -7$	◁ Multiplication Property of Equality (Case 2 only)
$x = 12$	**OR**	$x = -2$	◁ Addition Property of Equality

Check

3A EXPLORE

Example 2

Solve.
$2|x - 3| = 8$

This means that twice the distance between x and 3 is 8.

Plan

Implement

$2|x - 3| = 8$

$|x - 3| = 4$

Case 1:   Case 2:

                  OR

                  OR

                  OR

Explain

◂ Given

◂ Multiplication Property of Equality

◂ Definition of Absolute Value

◂ Multiplication Property of Equality (Case 2 only)

◂ Addition Property of Equality

☑ Checkpoint

Solve.

$2|x - 3| = 10$

▶ Graphing Solutions Using the Midpoint

- When absolute value equations are graphed, the _____

  and the _____ should be included on the number line.

- After you find the two solutions, determine the midpoint by finding the _____

  of the two solutions.

Example 3

Solve. Graph the midpoint and solutions.

$|2x - 3| = 6$

Plan Before solving for the two cases, first isolate the absolute value.
  · 2 (Multiply by 2)
  (−3) (Inside absolute value, subtract 3 (or add −3))
  | | (absolute value)

Implement

$|2x - 3| = 6$ ◀ Given

Case 1:   **Case 2:**

$2x - 3 = 6$ OR $-(2x - 3) = 6$ ◀ Definition of Absolute Value

              OR $2x - 3 = -6$ ◀ Multiplication Property of Equality (Case 2 only)

◀ Addition Property of Equality

◀ Multiplication Property of Equality

Explain

To determine the midpoint find the mean of the two solutions.

$\left(\frac{9}{2} + \left(-\frac{3}{2}\right)\right) \div 2$

$\frac{6}{2} \cdot \frac{1}{2} = \frac{3}{2}$

The midpoint of the solutions is $\frac{3}{2}$.

> Not all problems will require that you find the midpoint. However, the midpoint helps you see how absolute value solutions are the same distance from a common point.

☑ Checkpoint

Given the equation and solutions, determine the midpoint for the equation. Name the midpoint and graph the solutions.

$|6x - 2| + 1 = 5$

$x = 1 \text{ or } x = -\frac{1}{3}$

3A EXPLORE

▶ Writing Absolute Value Equations

- |x − (midpoint)| = (distance)

- When writing an absolute value equation, it is helpful to know the _____.

- In an absolute value equation, the midpoint is the number inside the _____ bars (when the coefficient of the variable is 1).

- In an absolute value equation, the _____ is what the absolute value equation is equal to.

Example 4

Jim and Cory have a difference of three Liberty nickels. If Jim collected 5 Liberty nickels, then how many Liberty nickels did Cory collect?

Plan Define variables.
  Write and solve an equation.

Solve the equation.
Let c = number of Liberty nickels Cory collected.

Implement

$|c − 5| = 3$

  OR

  OR

Explain

◀ Definition of Absolute Value

◀ Multiplication Property of Equality (Case 2 only)

◀ Addition Property of Equality

Explain
These solutions mean that Cory either has _____.

☑ Checkpoint

A package of granola can have a difference of 4 grams from the listed weight on the box. If the weight on the granola box says 510 grams, what is the lowest allowed weight of the box? What is the highest allowed weight?

Write an equation and solve.

Practice 1

Complete the problems on a separate sheet of paper.

Solve. Remember to check your work.

1) $\left|\frac{3}{4}x\right| = 5$

2) $|5x + 7| = 22$

3) $|3x - 6| = 4$

4) $4 = 5|v - 5|$

5) $2|k - 2| = 8$

6) $\left|\frac{2}{3}x - 1\right| = 10$

Graph the solutions and note the midpoint on a number line.

7) $3 = |x + 10|$

8) $\left|\frac{3}{2}k\right| = 9$

9) $4|w + 6| = 8$

10) $|2x + 6| = 9$

Solve. Remember to check your work.

11) The weight of a dozen large eggs must be within 1.5 oz of 25.5 oz. Write and solve an equation to represent the maximum and minimum weight allowed per dozen large eggs.

12) Malek scored an average of 9.5 points per game last season. This season, Malek has been 1.5 points above and below this average. Write and solve an equation to represent the fewest and the most points Malek has had this season.

3A MASTERY CHECK

✎ Mastery Check

✐ Show What You Know

Reilly's Tackle and Fishing Rentals was holding a rockfish fishing contest. The first rule stated that the rockfish caught must be 12.25 pounds. The second rule of the contest was that the rockfish must be 15.75 inches (plus or minus 1.25 inches).

Write an equation to show the shortest length and the longest length of the rockfish. Define your variable.

A) Write an equation to show the shortest length and the longest length of fish. Define your variable.

B) Solve your equation from Part A.

C) Write a formula that will work for any length of rockfish but still uses plus or minus 1.25.

·ılıı· Say What You Know

In your own words, talk about what you have learned using the objectives for this part of the lesson and your work on this page.

Practice 2

Complete the problems on a separate sheet of paper.

Solve. Remember to check your work.

1) $|-d| = 2$

2) $\left|\frac{1}{2}x + 6\right| = 3$

3) $|2r - 7| = 3$

4) $3|x - 6| = 4$

5) $4|z + 5| = 16$

6) $\left|\frac{5}{4}x + 3\right| = 0$

Graph the solutions and note the midpoint on a number line.

7) $6 = \left|\frac{3}{4}x\right|$

8) $|x - (-1)| = 1$

9) $|-2x - 1| = 5$

10) $|4r + 3| = 7$

Solve. Remember to check your work.

11) A group of anglers were catching fish. The weight of the fish must be within 2 pounds of the standard weight to be sold at market. If the standard weight is 23 pounds, what is the minimum and maximum the fish can weigh and still be sold?

12) Susanna was playing in a golf tournament. Par for the course is 65. She was 7 strokes from par. What were Susanna's possible scores?

3B EXPLORE

Part B: Multi-Step Absolute Value Equations

Objectives

In this part of the lesson, you will learn about multi-step absolute value equations.

By the end of this lesson, you will be able to do the following:

- ✓ Solve multi-step absolute value equations.
- ✓ Graph multi-step absolute value equation solutions on a number line.
- ✓ Determine when an absolute value equation has no solution or is an identity.

Why?

Understanding the different types of solutions to multi-step absolute value equations is important to determining if your solutions make sense.

Warm Up

1) What does it mean when the solution to an equation is all real numbers?

2) What does it mean for an equation to have no solutions?

Multi-Step Absolute Value Equations

- Solving multi-step absolute value equations is similar to solving equations without _____.

- Solving multi-step absolute value equations involves isolating the absolute value bars and then the _____.

Example 1

Solve for x.
$3|2x - 4| = 12$

Plan Determine the operations occurring to the variable.

- $\cdot 2$ (inside)
- -4 (inside)
- $|\ |$ (absolute value)
- $\cdot 3$ (outside)

To solve for x, follow the operations in backwards order using the inverse operations.

Implement   **Explain**

$3|2x - 4| = 12$ ◂ Given

$|2x - 4| = 4$ ◂ Multiplication Property of Equality

Case 1:   **Case 2:**

$2x - 4 = 4$ **OR** $-(2x - 4) = 4$ ◂ Definition of Absolute Value

                           OR $2x - 4 = -4$ ◂ Multiplication Property of Equality (Case 2 only)

$2x = 8$ **OR** $2x = 0$ ◂ Addition Property of Equality

$x = 4$ **OR** $x = 0$ ◂ Multiplication Property of Equality

Check

Case 1:   **Case 2:**

$x = 4$ $x = 0$

$3|2(4) - 4| = 12$ $3|2(0) - 4| = 12$

$3|8 - 4| = 12$ $3|0 - 4| = 12$

$3|4| = 12$ $3|-4| = 12$

$3(4) = 12$ $3(4) = 12$

$12 = 12$ ✓ $12 = 12$ ✓

3B EXPLORE

Example 2

Solve.

$11 - 2|x + 6| = -51$

Plan Determine what is happening to the variable. Then work backwards.

Implement   **Explain**

> It is important to remember when solving that the absolute value bars must be alone on one side of the equation before the problem can be split into Case 1 and Case 2.

Check

☑ Checkpoint

Say or write the steps occurring to the variable. Solve.

$5\left|\frac{1}{5}x - 2\right| + 3 = 23$

Absolute Value Equations with No Solution and All Real Numbers

- The absolute value of a number will always be greater than or equal to zero because _____ is a _____ number. It is not possible to travel a negative number of units.

- It is important to make sure the absolute value expressions are equal to a _____ number before solving for the two cases.

- Absolute value expressions that are set equal to a negative number have _____.

Example 3

Solve.
$-3|x+4| = 15$

◀ Multiplication Property of Equality

$|x+4| = -5$

- When any value will make an absolute value equation true, the solution is _____.

Example 4

Solve. Graph the solution(s) on the number line.

$|3x - 3x| = 5(2-2)$

Implement

$|3x - 3x| = 5(2-2)$

$|0| = 5(0)$

$0 = 0$

all real numbers, $\mathbb{R}$

Explain

◀ Given

◀ Simplify expressions

◀ This absolute value equation does not need to be separated into Case 1 and Case 2 because zero is the opposite of zero.

☑ Checkpoint

Solve.

$\frac{1}{2}|x-1| + 5 = 1$

Practice 1

Complete the problems on a separate sheet of paper.

Solve.

1) $|3x - 7| + 5 = 11$

2) $-\left|\frac{4}{5}x + 8\right| = -1$

3) $\frac{1}{2}|3x - 4| + 5 = 6$

Graph the solutions and note the midpoint on a number line.

4) $-2\left|\frac{3}{2}x - 5 - \frac{1}{2}x\right| = -8$

5) $-\frac{1}{2}|q + 8 + 3q - 2| = -6$

6) The average height of men in the Coes family is 5 feet 11 inches. The Coes have two sons within 3 inches of the family average for the men. What is the shortest possible height of the sons? The tallest possible height? Write an equation, solve, and graph.

Determine if the variables in the equations below have no solution, some (one or two) solutions, or is an identity.

7) $|x| = 0$

8) $|2x| = |-2x|$

9) $|x| = -1$

10) $4|z + 5| + 2 = 18$

11) $12 = -5|y| + 7$

12) $-|x + 15| = 26$

13) $3|7 - 2x + 2x - 3| - 1 = 11$

Explain why the given equation is an identity.

14) $|2x + 4 - 2x| = 4$

Mastery Check

Show What You Know

Complete each part of the problem.

A) Solve.

$\frac{4}{3}(2x-6) + 16 = 0$

B) Solve.

$\frac{4}{3}|2x-6| + 16 = 0$

C) How does changing the grouping symbols in Part A ( ) to absolute bars | | in Part B affect the solution(s)? Explain.

Say What You Know

In your own words, talk about what you have learned using the objectives for this part of the lesson and your work on this page.

Practice 2

Complete the problems on a separate sheet of paper.

Solve.

1) $-\frac{2}{3}\left|\frac{1}{4}x + 2\right| = -2$

2) $-|8x - 3| = -13$

3) $\frac{1}{10}|9 - 2x| - 3 = 0$

Solve. Draw a graph that includes the solutions and the midpoint.

4) $|2x - 3 - 6x + 1| = 2$

5) $-\frac{3}{2}|5 - 3x| + 1 = -2$

6) The average time to read a 300-page novel in Mr. Webber's class is 4 hours and 35 minutes. If all of the students are within 23 minutes of the average, what is the shortest time to read the novel? The longest time? Write an equation, solve, and graph on a number line.

Determine if the variables in the equations below have no solution, some (one or two) solutions, or is an identity.

7) $|2x - 2x| = 0$

8) $-\left|\frac{1}{3}x\right| = -2$

9) $|x - x - 12| = 11$

10) $9|2q - 2q| - 4 = -4$

11) $3|2x + 3 - x| + 4 = 3 - 2$

12) $|2r - 7| = 5$

13) $-\frac{1}{2}|q + 6 + 3q| + 3 = 6$

Explain why the given equation has no solution.

14) $|5x + 3| = -12$

TARGETED REVIEW 3

Targeted Review

In the Targeted Review, you will practice topics you have mastered in earlier lessons. Reviewing these concepts will help you be successful as you work through this unit.

Complete the problems on a separate sheet of paper.

Solve and write if the equation has one solution, no solution, or a solution of all real numbers.

1) $25 = 9 - 8(x - 9)$

2) $\frac{2}{3}(x + 9) = 3x - (2x - 6)$

3) $\frac{1}{2}(12x - 8) = 6x + 3$

4) $\frac{1}{4}x + 5 = \frac{2}{5}x - 1$

5) Solve for the indicated variable.
$V = \frac{lwh}{3}; w$

6) Briley bought three bottles of water and a tube of lip balm. The lip balm cost $2.00. If Briley paid $9.05 for the water and lip balm, how much did each bottle of water (w) cost? Write and solve the equation.

Evaluate. Then name all sets of numbers to which the value belongs.

7) $\frac{0.3}{0.05}$

8) $-\sqrt{16}$

9) Write the values from the set $\{-2, 0, 2, 15\}$ that are solutions for the inequality $x \leq 2$.

10) Use your formula sheet to find the value of the expression 3^0.

Multiple Choice

_____ 11) Which irrational number when squared will still be an irrational number?

A) π

B) $\sqrt{5}$

C) $\frac{\sqrt{3}}{2}$

D) $\frac{\sqrt{7}}{\sqrt{3}}$

_____ 12) Determine the number of solutions to the given equation.
$8x - 6x + 3 = \frac{1}{4}(8x + 12)$

A) no solution

B) one solution

C) two solutions

D) all real numbers

Lesson 4
Solving Inequalities

Outline

Part A
Single-Variable Inequalities

- Inequality Symbols and Wording
- Solving Inequalities
- Multiplying Inequalities by Negatives

Part B
Compound Inequalities

- Compound Inequalities with Two Symbols
- Absolute Value Inequalities
- Absolute Value Inequalities with No Solution or All Real Numbers

Targeted Review

Vocabulary

- inequality
- compound inequality
- absolute value inequality

Part A: Single-Variable Inequalities

Objectives

In this part of the lesson, you will learn about single-variable inequalities.

By the end of this lesson, you will be able to do the following:

- ⊘ Solve inequalities that include rational coefficients.
- ⊘ Graph solutions for an inequality on a number line.
- ⊘ Explain why the inequality symbol changes when multiplying by a negative factor.

Why?

Imagine you are traveling by plane with strict luggage weight limits. After packing your necessities, how many books can you carry with you? You can figure that out by solving for solutions to an inequality.

Warm Up

1) Name one possible solution for each of the following.

$a < 4$ _____

$b - 3 = 5$ _____

$2 \leq c$ _____

2) Name two variables from problem 1 that have more than one possible solution. Explain.

Inequality Symbols and Wording

- An _____ is a comparison of two expressions that are not equal.

Symbol	Represented Graphically	Symbol Name	Additional Wording
	open point	is not equal to	
	open point	is greater than	is more than >, is larger than >, is above, exceeds
	open point	is less than	is smaller than, is below
	closed point	is greater than or equal to	at least, has a minimum, is not smaller/less than
	closed point	is less than or equal to	at most, has a maximum, is not more/greater than, does not exceed

- The inequality symbols that are represented by an open point are _____.

- The inequality symbols that are represented by a closed point are _____.

- Write the inequality represented on the graph.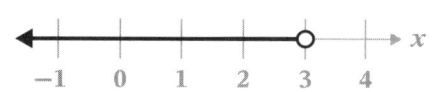

  _____

- If the variable switches sides, the _____ of the inequality must change.

- Write the inequality represented on the graph.

  _____

☑ Checkpoint

Write the symbol that best matches the situation. Will the point be open or closed when graphed?

A) Kyle needs at least $200 to purchase textbooks.

  t [ ] 200

B) Miranda can work a maximum of 10 hours this week.

  h [ ] 10

4A EXPLORE

▶ Solving Inequalities

- Equations and inequalities are solved using _____ methods.

Example 1

Compare the expressions $5w + 2 - 5$ and $2(w - 1) + 8$. Solve for the possible values of w in the comparison below.

$5w + 2 - 5 \;\square\; 2(w - 1) + 8$	◂ Given
$5w - 3 \;\square\; 2(w - 1) + 8$	◂ Combine like terms
$5w - 3 \;\square\; 2w - 2 + 8$	◂ Distributive Property (Multiply each term in $(w - 1)$ by 2)
$5w - 3 \;\square\; 2w + 6$	◂ Combine like terms
$3w - 3 \;\square\; 6$	◂ Addition Property of Equality (Add $-2w$ to expressions on both sides)
$3w \;\square\; 9$	◂ Addition Property of Equality (Add 3 to expressions on both sides)
$w \;\square\; 3$	◂ Multiplication Property of Equality (Multiply expressions on both sides by $\frac{1}{3}$)

Because equations and inequalities are solved using the same methods, no matter what symbol is used, the exact same steps will be used to solve for w. The only differences in these examples are the symbols and the solutions.

Equality: If $5w + 2 - 5 = 2(w - 1) + 8$, then $w = 3$.

Greater than: If $5w + 2 - 5 > 2(w - 1) + 8$, then $w > 3$.

Greater than or equal to: If $5w + 2 - 5 \geq 2(w - 1) + 8$, then $w \geq 3$.

Less than: If $5w + 2 - 5 < 2(w - 1) + 8$, then $w < 3$.

Less than or equal to: If $5w + 2 - 5 \leq 2(w - 1) + 8$, then $w \leq 3$.

> "Not equal to" (≠) is not shown because this symbol is primarily used when checking solutions.

EXPLORE 4A

☑ Checkpoint

Solve.

$3(x + 11) > -4$

▶ Multiplying Inequalities by Negatives

- When inequalities are multiplied or divided by a negative value, the _____ of the inequality symbol will change.

- When you multiply a negative number by a negative, the result is positive, which changes the _____ between the values.

- And so, when inequalities are multiplied or divided by a _____ value, the direction of the inequality symbol _____.

4A EXPLORE

Example 2

Solve. Graph the solution(s) on the number line.

$-\frac{1}{5}x + \frac{1}{3} \leq \frac{2}{3}$

Implement

$-\frac{1}{5}x + \frac{1}{3} \leq \frac{2}{3}$

$15\left(-\frac{1}{5}x + \frac{1}{3} \leq \frac{2}{3}\right)$

Explain

◀ Given

◀ Clear the fractions in the inequality first using the LCD (as you did in Lesson 2).

◀ Addition Property of Equality
(Isolate the variable by subtracting 5 from both sides)

◀ Multiplication Property of Equality
(Divide both sides by −3)

◀ The inequality symbol changes direction because the inequality was divided on both sides by a negative number.

Check

The two expressions will be equal at $x = -\frac{5}{3}$.

$-\frac{1}{5}\left(-\frac{5}{3}\right) + \frac{1}{3} \leq \frac{2}{3}$

$\frac{5}{15} + \frac{1}{3} \leq \frac{2}{3}$

$\frac{5}{15} + \frac{5}{15} \leq \frac{10}{15}$

$\frac{10}{15} \leq \frac{10}{15}$ ✓

Values where $x < -\frac{5}{3}$ will result in false inequalities when substituted into the original.

$-\frac{1}{5}(-5) + \frac{1}{3} < \frac{2}{3}$

$1 + \frac{1}{3} < \frac{2}{3}$

$\frac{4}{3} < \frac{2}{3}$ ✗

Solutions where $x > -\frac{5}{3}$ will result in true inequalities when substituted into the original. This means the shading belongs to the right of the boundary point.

$-\frac{1}{5}(0) + \frac{1}{3} \leq \frac{2}{3}$

$0 + \frac{1}{3} \leq \frac{2}{3}$

$\frac{1}{3} \leq \frac{2}{3}$ ✓

☑ Checkpoint

Solve. Graph the solution(s) on a number line.

$-\frac{2}{3}x + 8 > -2$

Practice 1

Complete the problems on a separate sheet of paper.

Write the inequality sign that matches the graph. From the set of numbers, $\left\{-7, -3, -1.2, 0, \frac{3}{4}, 1, 2\right\}$, name any that are solutions for the variable.

1) $p \;\square\; 1$

2) $w \;\square\; 2$

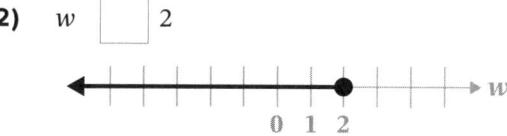

Graph the solutions to the given inequalities on a number line. From the set of numbers, $\left\{-7, -3, -1.2, 0, \frac{3}{4}, 1, 2\right\}$, name any that are solutions for the variable.

3) $-r \leq 3$

4) $q < 0$

Solve. Graph solution(s) on a number line. Justify your steps.

5) $\frac{1}{2}w + 3 \geq 1$

6) $\frac{3}{8}(x - 5) < \frac{1}{5}$

7) Russell has already saved $110. If he saves $30 per week, how many weeks must he save to have at least $500? Write and solve an inequality. Remember to define your variable.

8) Adia rented a car for her trip. She could drive *at most* 250 miles without being charged an additional fee. Adia traveled twice as far on Thursday as she did on Friday. On Saturday she traveled 70 miles before returning the car. How far could Adia have traveled on Friday without being charged the additional fee? Write and solve an inequality. Remember to define your variable.

Solve and graph each problem on a number line.

9) $-3x + 1 > 4$

10) $3x + 1 > 4$

11) What are the similarities and differences between problems 9 and 10? Explain.

12) $2n + 1 - 6n - 4 \geq 3n - 7$

13) $-3x + 2 < -1$

14) $-5(j - 2) \leq -10$

15) $-6 - 8x < -10x + 3$

16) The quotient of a number and negative five, plus eight is less than or equal to negative three. Write and solve the inequality.

4A MASTERY CHECK

✅ Mastery Check

✏️ Show What You Know

Using the integers −4 to 4 *only once*, fill in the boxes so that you create the following solutions.

−4 −3 −2 −1 0 1 2 3 4

A) An inequality has no solution when the variable(s) simplify out of the inequality and the numbers remaining create an untrue statement. (e.g., 8 < 3). Create an inequality that has *no solution*.

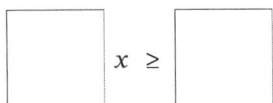

B) x is greater than or equal to a *negative number*.

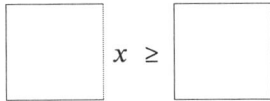

C) x is less than or equal to a *positive number*.

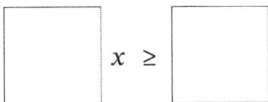

D) Explain your thinking.

🔊 Say What You Know

In your own words, talk about what you have learned using the objectives for this part of the lesson and your work on this page.

Practice 2

Complete the problems on a separate sheet of paper.

Write the inequality sign that matches the graph. From the set of numbers, $\left\{-7, -3, -1.2, 0, \frac{3}{4}, 1, 2\right\}$, name any that are solutions for the variable.

1) $c \;\boxed{}\; -2$

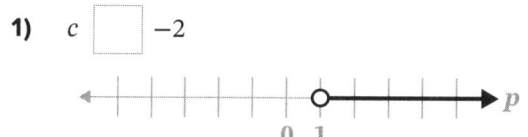

2) $n \;\boxed{}\; -1$

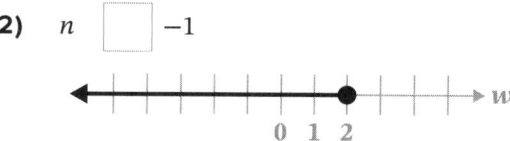

Graph the solutions to the given inequalities on a number line. From the set of numbers, $\left\{-7, -3, -1.2, 0, \frac{3}{4}, 1, 2\right\}$, name any that are solutions for the variable.

3) $e > 0$

4) $-k \geq 0$

Solve. Graph the solution(s) on a number line. Justify your steps.

5) $0.2c - 0.3 > 0.4$

6) $2x + \frac{1}{2} > \frac{7}{3}$

Write and solve an inequality. Remember to define your variable.

7) Branson already saved $800 toward the purchase of a used car. Branson was paid $10 for each hour worked. If Branson needs a minimum of $3,250 to purchase the car, how many hours will he need to work?

8) Jackson is going camping. His backpack can hold no more than 35 pounds. His tent is 4 pounds, his sleeping bag is $1\frac{1}{2}$ pounds, his clothes weigh $8\frac{1}{2}$ pounds, and his cooking supplies weigh 3 pounds. Each package of food weighs 2 pounds. How many packages of food can he bring without exceeding the limit?

Graph the solutions on a number line.

9) $q > -5$

10) $-q < 5$

11) Are the graphs the same? Why or why not?

Solve. Graph the solutions on a number line.

12) $2(u + 7) + 1 > 3(u - 4) + 2$

13) $-6(r - 3) \leq 24$

14) Find the error, write it down, and then correct it. Explain and provide the correct solution.

$$8g + 2 \geq -5g - 10$$
$$8g + 2 \geq -10$$
$$8g \geq -12$$
$$g \leq -\frac{3}{2}$$

Solve the inequality.

15) $-6x + 11 < -(4x + 3)$

16) Three times a number plus ten, decreased by seven times the same number is at least twelve. Write and solve the inequality.

4B EXPLORE

Part B: Compound Inequalities

Objectives

In this part of the lesson, you will learn about compound inequalities.

By the end of this lesson, you will be able to do the following:
- Solve and graph solutions for single-variable compound inequalities.
- Solve and graph solutions for single-variable inequalities that contain absolute value.
- Identify inequalities containing absolute value as having no solution or as true for all real numbers.

Why?

Imagine you are given an assignment for government studies that requires you to poll 95 people but can vary by up to 12. Knowing how to solve compound inequalities would give you the range of the number of people you need to poll to complete the assignment.

Warm Up

Fill in the letter from the list provided that matches the equation with the correct number of solutions.

A) no solutions   B) one solution   C) two solutions   D) all real numbers

_____ 1) $x + 4 = 6$

_____ 2) $|x - 5| = 7$

_____ 3) $x = x$

_____ 4) $|x + 8| = -2$

Compound Inequalities with Two Symbols

- Compound inequalities use _____ or _____ to combine the solutions of two inequalities for graphing on one number line.

- Inequalities that use the word OR have solutions that are true for _____ of the inequalities.

- Solutions to inequalities that use the word AND must be true for _____ inequalities.

    - When there is no number that makes both inequalities true for an AND problem, _____ exists.

- The solutions to a compound inequality should be graphed on _____ number line.

EXPLORE 4B

Example 1

Solve. Graph the solution(s) on the number line.

$x + 7 > 5$ OR $-2x - 6 \geq 8$

$x > -2$ 　　　　$-2x \geq 14$ ◀ Addition Property of Equality

　　　　　　　　$x \leq -7$ ◀ Multiplication Property of Equality

$x > -2$ OR $x \leq -7$

Example 2

Solve. Graph the solution(s) on the number line.

$3x - 2 < 14 - x < 5x$

Plan Write problem as 2 inequalities with AND between them.
Solve both inequalities.
Graph all solutions.

$3x - 2 < 14 - x$　　　　**AND**　　　　$14 - x < 5x$

> When writing as one compound inequality statement, the smaller value will be on the left and the larger value on the right, like that of a number line.

☑ Checkpoint

Solve. Graph the solution on the number line.

$-8 \leq 2y - 8 < 4$

Algebra 1 Student Worktext　　Lesson 4: Solving Inequalities > Part B: Compound Inequalities > Explore　71

4B EXPLORE

▶ Absolute Value Inequalities

- With absolute value inequalities the direction of the _____ determines whether it is an AND or OR compound inequality.

- As with absolute value equations, absolute value inequalities require _____ case(s) to find the complete solution.

- Absolute value expressions that are greater than or equal to a constant create an _____ compound inequality.

- Absolute value expressions that are less than or equal to a constant create a _____ compound inequality.

Example 3

Write and solve an inequality.
Graph the solution(s) on the number line.

The absolute value of a number x and four is less than three.
$|x - 4| < 3$

Case 1:
$x - 4 < 3$

AND

Case 2:
$-(x - 4) \,\fbox{$<$}\, 3$

> This inequality is open away from the absolute value expression. This will be a AND compound inequality.

EXPLORE 4B

Example 4

Write and solve an inequality. Graph the solution(s) on the number line.

7 added to the absolute value of 2v minus 5 plus 2 is greater than or equal to 14.

$|2v - 5 + 2| + 7 \geq 14$

$|2v - 3| + 7 \geq 14$ ◀ Combine like terms

$|2v - 3| \geq 7$ ◀ Addition Property of Equality

Case 1:   **Case 2:**

◀ Definition of absolute value (OR is used because $|\ | \geq 7$)

◀ Multiplication Property of Equality (Case 2)

◀ Addition Property of Equality

◀ Multiplication Property of Equality

> If you cannot recall whether the inequality will be AND or OR, graph the solutions first, then go back and write in the word AND or OR.

☑ Checkpoint

Solve. Graph solution(s) on the number line.

$|3x + 4| - 10 > 2$

4B EXPLORE

▶ Absolute Value Inequalities with No Solution or All Real Numbers

- Because absolute value represents a _____, a true solution must be greater than or equal to zero.

- An absolute value inequality has _____ if the constant is less than zero before writing Case 1 and Case 2.

- An absolute value inequality has a solution of _____ if the absolute value expression is greater than or equal to zero or any negative value.

Example 5

Solve.

$|x + 12| \leq -4$

no solution — This inequality has no solution because $|x + 12|$ will always result in a value that is greater than or equal to zero.

> Although $|x + 12| < 0$ has no solution, $|x + 12| \leq 0$ has one solution, $x = -12$.

Example 6

Solve.

$|x + 8| > -2$

all real numbers — The solution to this inequality is all real numbers since the absolute value will always be greater than or equal to zero.

> Although $|x + 8| \geq 0$, is true for all values of x, $|x + 8| > 0$ is not true when $x = -8$, because $|x + 8|$ would be equal to zero.

☑ Checkpoint

Explain why the absolute value inequality has no solutions or is true for all real numbers.

A) $|4x - 8| < 0$

B) $\left|\frac{1}{2}x + 6\right| > -\frac{2}{3}$

Practice 1

Complete the problems on a separate sheet of paper.

Graph the solutions to the given inequalities on a number line. Name all of the values that are solutions for the variable from the set $\left\{-8, -4, -1.2, -\frac{3}{4}, 0, 0.5, \frac{10}{3}, 5, 10\right\}$.

1) $n > -1.2$

2) $n \leq \frac{10}{3}$

3) $n > -1.2$ AND $n \leq \frac{10}{3}$

4) $n > -1.2$ OR $n \leq \frac{10}{3}$

5) Write the compound inequality, using AND or OR, that represents the given graph.

Solve the following inequalities and graph the solutions on a number line.

6) $-1 < 5x + 9 \leq 19$

7) $-4x + 1 \leq 5$ OR $-3x + 2 < -4$

8) Draw a graph to represent the solutions when problem 7 uses AND.

9) $0 \leq 6c + 12 \leq 6$

10) Name 3 values that make compound inequality true in problem 9.

11) $2m > 14$ OR $-2m > 10$

12) How would the solution to problem 11 change if it was an AND statement?

Solve. Graph the solution(s) on a number line.

13) $|4x - 3| > 4$

14) $\frac{1}{2}|h + 1| - 4 \leq 2$

15) What would a graph look like if the equation was $\frac{1}{2}|h + 1| - 4 \geq 2$?

16) Solve. Justify your steps with properties. $-5|p| - 3 > -3$

17) Explain how you would graph your solution(s) to problem 16.

18) Write an inequality with an absolute value where the solution is all real numbers.

19) Using the same values as in Problem 18, write an inequality that has no solution.

20) While practicing for her driving exam, Melina realized her cruise control always stays within 2 miles per hour of the set speed. If Melina set the cruise control to 35 miles per hour, what is the range of speeds her car could actually go? Write and solve the compound inequality. Remember to define your variable.

21) Abul has to write an essay that is between 400 and 500 words. He has already written 150 words. How many more words can Abul write and remain within the word limit? Write and solve the compound inequality. Remember to define your variable.

4B MASTERY CHECK

Mastery Check

Show What You Know

Dan is training for a marathon.

A) Write a compound inequality for the range in hours that Dan could train in one day. Explain your thinking.

B) Suppose Dan never trains for more than 8 hours a day. Rewrite your compound inequality with this information.

C) The marathon rules state that all runners must finish the race within 2.5 hours of the median time of 4 hours. Write an absolute value inequality and solve to find the range in finishing times.

Say What You Know

In your own words, talk about what you have learned using the objectives for this part of the lesson and your work on this page.

Practice 2

Complete the problems on a separate sheet of paper.

Solve. Graph all possible solutions on a number line.

1) Given $p < -5, p > 1$ graph the inequalities individually and the different compound inequalities of AND and OR.

2) Given the graph, write the compound inequality.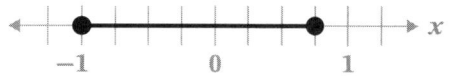

3) Solve. Graph all possible solutions on a number line.

4) $5 < \frac{2}{3}x + 2 < 10$

5) $-3x + 1 > 10$ OR $2x - 3 > -5$

6) $3 \leq -4x - 1 < 15$

7) $4a < 12$ OR $7a > 21$

Solve. Graph all possible solutions on a number line.

7) $\left|\frac{5}{3}c + 9\right| \leq 6$

8) How would the inequality to the left need to change so that the solution had an OR statement?

9) $2|c - 2| + 3 > 11$

10) $\left|\frac{3}{7}k\right| < -18$

Find the error, write it down, then correct it. Explain your answer.

11)
$|2x| + 5 \geq 5$
$|2x| \geq 0$

$2x \geq 0$ OR $-(2x) \leq 0$
OR $2x \leq 0$
$|2x| \geq 0$ OR $-(2x) \geq 0$

12)
$2|y - 3| + 5 \geq -3$
$2|y - 3| \geq -8$
$|y - 3| \geq -4$

$y - 3 \geq -4$ OR $-(y - 3) \geq -4$
OR $y - 3 \geq 4$
$y \geq -1$ OR $y \geq 7$

13) At a local school, the standard class size is 25 students. This can vary by up to 8 students. What is the range of students that can be in a class? Write and solve a compound inequality with a defined variable.

TARGETED REVIEW 4

Targeted Review

In the Targeted Review, you will practice topics you have mastered in earlier lessons. Reviewing these concepts will help you be successful as you work through this unit.

Complete the problems on a separate sheet of paper.

Solve. Write if the equation has no solution, one or two solution(s), or a solution of all real numbers.

1) $|2h - 3| = 5$

2) $-3(x + 3) = 2x - 3 - (x + 4)$

Write the inequality symbol that demonstrates the relationship.

3) $(-8)^2$ ☐ -8^2

4) Why are the answers to the two expressions in problem 3 different?

Name the property being demonstrated using your formula sheet.

5) D'von was checking a solution. He found that $12 = 12$. What algebraic property does this demonstrate?

6) What property allows you to multiply by the reciprocal so the result is one?

Solve.

7) Ida received notice that her checking account was overdrawn by $20. (This means her account balance says −$20.) The bank charged her a $40 fee for overdrawing the account. Then Ida deposited $35. The next day she deposited twice that amount. Then she deposited a $75 check but also wrote a check for $22.50. What is the balance of her checking account?

8) One-half a number, plus seven is equal to three times the difference of the same number and six. Write and solve an equation.

9) Use your formula sheet to determine how many centimeters are in 1 inch.

10) Use your formula sheet to write the formal for a linear equation in slope-intercept form.

Multiple Choice

_____ 11) Select the graph that represents the correct solutions for the equation below.
$|x - 2| = 6$

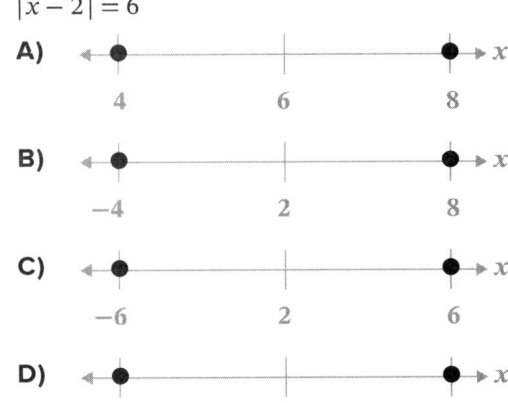

12) Select two numbers whose product is *irrational*.

☐ $\sqrt{3}$

☐ $\sqrt{27}$

☐ -9

☐ $\frac{2}{9}$

Lesson 5
Ratios, Proportions, and Rates

Outline

Part A
Ratios and Proportions
- Equivalent Ratios
- Solving Proportions

Part B
Unit Conversions
- Single Unit Conversions
- Multiple Unit Conversions
- Compound Unit Conversions

Targeted Review

Vocabulary

- ratio
- proportion
- cross product
- unit conversion (dimensional analysis)
- unit multiplier

5A EXPLORE

Part A: Ratios and Proportions

Objectives

In this part of the lesson, you will learn about ratios and proportions.

By the end of this lesson, you will be able to do the following:

- Determine equivalent ratios.
- Solve a proportion using cross products.

Why?

Which brand and quantity of sports drink is the best buy at your local grocery store? You can figure out the answer to that and many other everyday situations using ratios and proportions.

☁ Warm Up

Rewrite all equations with integer coefficients, then solve.

1) $\frac{2}{5}n - \frac{1}{3} = \frac{3}{2}$

2) $\frac{3}{8}(x - 4) = \frac{3}{4}$

▷ Equivalent Ratios

- Three ways that ratios can be written using the variables a and b are:

- The _____ in which a ratio is written is very important.

- Ratios compare:

    - _____

    - _____

    - _____

- _____ ratios are equal to one another and can be written as a proportion.

- To show ratios are equivalent, simplify to _____, or find a _____.

Example 1

Determine which, if any, of the ratios are equivalent to $\frac{3}{4}$.

9:12   8 to 6

Plan Write the ratios as fractions. Compare them to $\frac{3}{4}$

Implement

Compare $\frac{3}{4}$ and $\frac{9}{12}$

Multiply by 1 $\left(\text{or } \frac{3}{3}\right)$:

$$\frac{3}{4} \cdot \frac{3}{3} = \frac{9}{12}$$

$$\frac{3}{4} = \frac{3}{4}$$

These are equivalent ratios.

Compare $\frac{3}{4}$ and $\frac{8}{6}$

Divide by 1:

$$\frac{8 \div 2}{6 \div 2} = \frac{4}{3}$$

$$\frac{3}{4} \neq \frac{4}{3}$$

These are not equivalent ratios.

Explain

9:12 is equivalent to $\frac{3}{4}$.

8 to 6 is not equivalent to $\frac{3}{4}$.

☑ Checkpoint

Select all ratios that are equivalent to 5:8.

☐ 15 to 24

☐ 10:13

☐ $\frac{10}{16}$

☐ 2:5

5A EXPLORE

▶ Solving Proportions

- A _____ is an equation that sets two ratios equal to one another.

- Proportions can be solved by finding the _____ of ratios.

- For a proportion to be true, the cross products must be _____ to one another.

- The cross product is found by multiplying:

  - the _____ of the first ratio by the _____ of the second ratio.

  - the _____ of the first ratio by the _____ of the second ratio.

- If more than one term is in the numerator or denominator of a ratio, use the _____ Property to solve the proportion.

Example 2

Solve.

$\frac{2x+1}{4} = \frac{3x-5}{5}$ **Plan** Find the cross product.

Implement   **Explain**

$(2x+1)(5) = (4)(3x-5)$ ◁ Cross product

$10x + 5 = 12x - 20$ ◁ Distributive Property

◁ Addition Property of Equality

◁ Multiplication Property of Equality

☑ Checkpoint

Solve.

$\frac{4}{5} = \frac{x+6}{10}$

Practice 1

Complete the problems on a separate sheet of paper.

Mr. Blake's music class had 15 flute players and 10 clarinet players.

1) Write an equivalent ratio of clarinet players to flute players in simplest terms.

2) Write an equivalent ratio of flute players to clarinet players in simplest terms.

3) Are the ratios the same in problems 1 and 2? Explain.

4) Create a ratio in simplest form that compares the number of flute players to the total number of people in the class.

---

5) Write all of the ratios below that are equivalent to 2:3.

   8:12   9:6   30:20   10:15

6) Write and solve a proportion for the given figure.   7) Solve the proportion: $\frac{5}{a} = \frac{20}{11}$

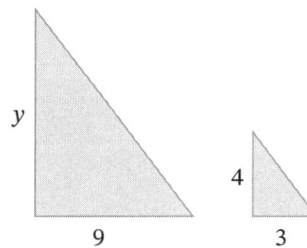

8) The ratio of apples to grapefruits used in the salad was six to five. If twelve apples were used, how many grapefruits were used?

9) There are 30 days in the month of September. If the ratio of cloudy to sunny days was one to two, how many days were cloudy and how many were sunny?

Solve.

10) $\frac{2x+1}{7} = \frac{4}{9}$ 11) $\frac{8}{x-2} = \frac{9}{x+3}$ 12) $\frac{x+15}{35} = \frac{x}{20}$ 13) $\frac{2x}{x+1} = \frac{5}{8}$

14) Six yards of fabric cost $76.50. How many yards of the same fabric can be purchased for $134? Round to the hundredth.

15) A pizza party for a classroom of 25 students cost $63 for 7 pizzas. What would the cost be for 150 students with a $20 coupon?

5A MASTERY CHECK

📋 Mastery Check

✏️ Show What You Know

The bakers at the Pancake Palace say the perfect pancake recipe uses the following ratio:

$$2 \text{ parts flour}: 2 \text{ parts liquid}: 1 \text{ part egg}: \frac{1}{2} \text{ part fat}$$

When this recipe is made using cups as the measure, one batch will yield 12 large pancakes.

Oliver works at the Pancake Palace. He needs to have 54 pancakes ready to make when the doors open.

A) What is the number of times the pancake recipe ratio needs to be multiplied by? Show your work.

B) Determine how much of each ingredient Oliver will need to make enough pancakes.

C) If the original recipe calls for 2 teaspoons of baking powder, how many *tablespoons* will be added to the batch to make 54 pancakes?

🔊 Say What You Know

In your own words, talk about what you have learned using the objectives for this part of the lesson and your work on this page.

Practice 2

Complete the problems on a separate sheet of paper.

At the 3R Farm, there are 16 cows for every 56 chickens.

1) Write a ratio of chickens to cows in simplest form.

2) Write the ratio of cows to chickens in simplest form.

3) Write a ratio comparing the number of cows to the total number of animals in simplest form.

The animal shelter held a pet adoption day. For every 5 dogs adopted, 2 cats were adopted.

4) How many cats would have been adopted if 10 dogs were adopted?

5) What would have been the total number of adopted pets if 6 cats were adopted?

6) Write and solve a proportion to find the missing side of the similar rectangles.

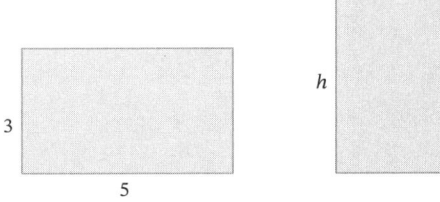

7) The ratio of time Mary spends on sports compared to school work is two to three. If she spends four hours a day practicing sports, how many hours does she spend studying?

8) The ratio of average snowfall in two towns is four to five. If the first town, Allentown, usually gets 22 inches a year, what is the average snowfall in the second town, Bakersfield?

9) Squirrels outnumbered rabbits by a ratio of eight to seven. If 56 rabbits were present, how many squirrels were there?

Solve.

10) $\frac{x-3}{5} = \frac{3}{10}$

11) $\frac{x}{18} = \frac{x+1}{100}$

12) $\frac{x}{30} = \frac{x}{400}$

13) $\frac{6-x}{3} = \frac{x}{5}$

14) Rob used a half gallon of paint to cover 250 square feet. How much paint would he need to cover 820 square feet? Round your answer to the whole gallon.

15) Bill can complete three fifths of a project in 30 minutes. How long will it take to complete the entire project?

Part B: Unit Conversions

Objectives

In this part of the lesson, you will learn about unit conversions.

By the end of this lesson, you will be able to do the following:

- Convert units for a value using a single conversion.
- Convert units for a value using multiple conversions.
- Convert compound units for a value.

Why?

Converting units is a skill that you will use often for cooking meals, building projects, and calculating distances and weights when exercising, just to name a few.

Warm Up

Use your Formula Sheet.

1) How many centimeters are in one meter?

2) What are two units of measurement that are equal to 36 inches?

Single Unit Conversions

- _____, or _____, is the process of converting one measurement to another.

- In order to compare two measurements, the unit of measurement must be the _____.

- To convert between units, you need to know the desired unit of measure and the correct _____.

- A unit multiplier is a _____ that is equal to 1, containing different units.

Example 1

The Holly family found that the perimeter of their garden was 24.5 yards in length. The lumber supply center sold fence pieces measured in feet. Convert the perimeter to feet to find how much fencing the Holly family needs to surround their garden.

Plan Conversion: 1 yd = 3 ft

The unit multiplier is either $\frac{1 \text{ yd}}{3 \text{ ft}}$ or $\frac{3 \text{ ft}}{1 \text{ yd}}$

Since the problem asks you to convert to feet, the correct unit multiplier is $\frac{3 \text{ ft}}{1 \text{ yd}}$

Implement

24.5 yd · $\frac{3 \text{ ft}}{1 \text{ yd}}$ =

24.5 · 3 ft = 73.5 ft

Explain

◄ Inverse Property (The unit "yard" simplifies out of the expression.)

Extra caution should be taken when converting _____ or _____ units.

Square Yards to Square Feet

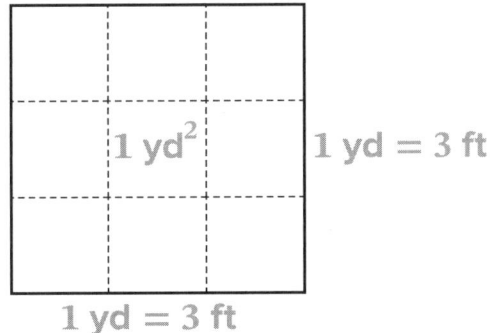

1 yd = 3 ft

1 yd = 3 ft

Example 2

Find the value of the conversion for the specified units below.
Round your answer to the nearest hundredth.

16.65 ft² = _____ yd²

Plan Conversion: 1 yd = 3 ft

Use the unit conversion $\frac{1 \text{ yd}}{3 \text{ ft}}$ twice because the units are squared.

Implement

$(16.65 \text{ ft}^2)\left(\left(\frac{1 \text{ yd}}{3 \text{ ft}}\right)\left(\frac{1 \text{ yd}}{3 \text{ ft}}\right)\right) =$

> A cubic unit can also be converted by using a single conversion three times.

5B EXPLORE

> ☑ **Checkpoint**
>
> Using your Formula Sheet, convert 100 inches into feet.
> Round to the nearest hundredth.

▶ Multiple Unit Conversions

- Many times using _____ unit multiplier is not enough to convert from one unit to another.

- The process of _____ is used until the desired conversion is reached.

Example 3

Summer made 3.75 gallons of punch for her sister's birthday party. Each punch glass (pg) will hold $\frac{3}{4}$ cup of punch. How many punch glasses can be filled?

Plan   3.75 gal = _____ pg

Conversions

$1 \text{ pg} = \frac{3}{4} \text{ c}$

$1 \text{ pt} = 2 \text{ c}$

$1 \text{ qt} = 2 \text{ pt}$

$1 \text{ gal} = 4 \text{ qt}$

Think about how the dimensional analysis should be set up so only "pg" for punch glass will remain.

Implement

$$(3.75 \text{ gal})\left(\frac{\text{qt}}{\text{gal}}\right)\left(\frac{\text{pt}}{\text{qt}}\right)\left(\frac{\text{c}}{\text{pt}}\right)\left(\frac{\text{pg}}{\text{c}}\right) = \_\_\_ \text{ pg}$$

$$(3.75 \text{ gal})\left(\frac{4 \text{ qt}}{1 \text{ gal}}\right)\left(\frac{2 \text{ pt}}{1 \text{ qt}}\right)\left(\frac{2 \text{ c}}{1 \text{ pt}}\right)\left(\frac{1 \text{ pg}}{\frac{3}{4} \text{ c}}\right) = \_\_\_ \text{ pg}$$

$$(3.75 \text{ g\cancel{al}})\left(\frac{4 \cancel{\text{qt}}}{1 \cancel{\text{gal}}}\right)\left(\frac{2 \cancel{\text{pt}}}{1 \cancel{\text{qt}}}\right)\left(\frac{2 \cancel{\text{c}}}{1 \cancel{\text{pt}}}\right)\left(\frac{1 \text{ pg}}{\frac{3}{4} \cancel{\text{c}}}\right) = \_\_\_ \text{ pg}$$

EXPLORE 5B

☑ Checkpoint

Convert 114.3 centimeters to yards.
List the unit conversions needed to evaluate using your Formula Sheet.

▶ Compound Unit Conversions

- Units of measure are often combined to make what is called a(n) _____.

- Miles per hour written as a unit multiplier is _____.

Example 4

Tristan could run at a speed of 3.4 miles per hour. How fast is this in feet per second?
Round your answer to the nearest hundredth.

Plan $3.4 \frac{mi}{hr} = $ _____ $\frac{ft}{sec}$

Conversions:

$5{,}280 \text{ ft} = 1 \text{ mi}$

$60 \text{ min} = 1 \text{ hr}$

$60 \text{ sec} = 1 \text{ min}$

Implement

$$\left(\frac{3.4 \text{ mi}}{1 \text{ hr}}\right)\left(\frac{5{,}280 \text{ ft}}{1 \text{ mi}}\right)\left(\frac{1 \text{ hr}}{60 \text{ min}}\right)\left(\frac{1 \text{ min}}{60 \text{ sec}}\right)$$

$$\frac{3.4 \, (5{,}280 \text{ ft}) \, (1) \, (1)}{(1) \, (1) \, (60) \, (60 \text{ sec})} =$$

Because there are two units in the final answer there will be one remaining unit in both the numerator and denominator when simplified.

☑ Checkpoint

A cheetah can run 90 feet per second. How fast is this in miles per hour?
Round to the nearest hundredth.

Practice 1

Complete the problems on a separate sheet of paper.

Use your Formula Sheet to determine the unit multiplier(s).

1) Convert the given units. 5 cm = _____ in

   A) Which unit of measure belongs in the numerator of the unit multiplier? Why?

   B) Which unit of measure belongs in the denominator of the unit multiplier? Why?

   C) Convert to the given unit. Round to the nearest unit.

Convert the given units.

2) 2 Tbsp = _____ tsp (Round to the nearest unit.)

3) 1.75×10^4 lb = _____ ton (1 ton = 2000 lb) (Provide the exact answer.)

4) Miranda wanted to make Turkish delight. The recipe calls for 10 milliliters of lemon juice. However, Miranda's measuring spoons were only in tablespoons. How many tablespoons of lemon juice does Miranda need for the recipe? Write the solution as a simplified fraction with no common factors between the numerator and denominator.

5) 0.8 ft^2 = _____ in^2

6) 750 cm^2 = _____ in^2 (Round to the nearest hundredth.)

7) Nick was trying to find the volume of a rectangular prism. The dimensions for the prism were given as 1.2 inches, 4.0 inches, and 1.5 feet. Nick's solution was as follows:
(1.2 in)(4.0 in)(1.5 ft) = 7.2 ft^2.
Find and correct Nick's error to provide a solution in cubic inches. Round to the nearest tenth.

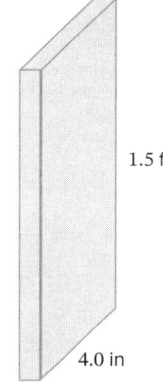

1.5 ft

4.0 in

1.2 in

Convert the given units. If necessary, round to the nearest hundredth.

8) 7.5 c = _____ qt

9) 5 day(s) = _____ sec

10) Craig was looking to buy 25 feet of fencing for his garden. However, he found that online the measurements were in meters. How many meters of fencing does Craig need for his garden?

11) Jamie has two job offers with two different companies. One company is offering to pay her $12.50 per hour. The other company is offering to pay her $0.52 per order for packaging orders. What is the average time, in minutes, that Jamie would need to fill each order to make the same amount of money at either job? (minutes per order) Round to the nearest tenth.

Use the table below for problems 12–14. Round your answer to the nearest hundredth.

	Alaska	Mongolia
Land Area	3.8 million mi^2	1.56 million km^2
Est. Population in 2020	732,000 people	3.2 million people
Precipitation in Capital	92.2 in/yr (Juneau)	378 mm/yr (Ulaanbaatar)

12) What is the difference in rainfall in inches between Alaska and Mongolia's capital cities? (1 in = 25.4 mm)

13) If the land area of Alaska was divided evenly among its population, how many square miles would each person in Alaska have?

14) If the land area of Mongolia was divided evenly among its population, how many square miles would each person in Mongolia have? (1 km^2 = 0.386 mi^2)

5B MASTERY CHECK

📋 Mastery Check

✏️ Show What You Know

Sandy City experienced a flood. The entire city was flooded with one inch of rain that was causing damage due to the weight of all the water.

A) The high school soccer field was 115 yards by 70 yards and had one inch of standing water. What is the volume of water on the field?

B) If water weighs 62 pounds per cubic foot, how much did the water on the soccer field weigh in tons?

C) The entire city has an area of 21 square miles. If the town is covered in one inch of water, what is the total weight of the flood waters in the city in pounds? What is the weight in tons?

🔊 Say What You Know

In your own words, talk about what you have learned using the objectives for this part of the lesson and your work on this page.

Practice 2

Complete the problems on a separate sheet of paper.

Convert the given units. Use your Formula Sheet to determine the unit multiplier(s).

1) 3.2 hr = _____ min.

   A) Which unit of measure belongs in the numerator of the unit multiplier? Why?

   B) Which unit of measure belongs in the denominator of the unit multiplier? Why?

   C) Convert to the given unit. Round to the nearest ten.

2) $\frac{3}{4}$ yd = _____ ft

3) 84 oz = _____ lb

4) Jason and Maddie need seven and a half pounds of flower petals for their project. However, flower petals are sold in ounces. How many ounces of flower petals do they need for their project? Round to the nearest unit.

5) 4.5 ft^2 = _____ yd^2

6) 1 mi^2 = _____ ft^2

7) Shayla's and Shane's swim times for 25 meters are listed below. Shayla wants to compare her time to Shane's. Convert Shayla's time to minutes in order to determine who swam 25 meters faster.

   Shayla's 25 meter time: 165 sec   Shane's 25 meter time: 2.8 min

   165 sec = _____ min

8) 115,200 min = _____ day(s)

9) 1 mi^2 = _____ acre(s) (1 acre = 43,560 ft^2)

10) Blake wants to print 2,000 buttons for his campaign on his 3-D printer. He can print 150 buttons from one spool of filament, and each spool costs $21. How much would it cost him to print the campaign buttons?

11) Tanya was getting paid $15 per hour working for 30 hours each week. How much would Tanya earn working 50 weeks in the year? Round to the nearest unit.

12) An ice hockey rink is laying down new ice. The dimensions of the rink are 200 feet by 85 feet, and the ice will be 1 inch thick when finished. If the weight of a cubic foot of ice is 57.2 pounds, what is the weight of the ice rink?

13) The rink manager said it will take 48 hours for the new ice rink in problem 11 to be completed. Each layer of ice is $\frac{1}{32}$ of an inch and must be layered on at even time intervals. How many minutes will be needed for each layer to freeze?

TARGETED REVIEW 5

Targeted Review

In the Targeted Review, you will practice topics you have mastered in earlier lessons. Reviewing these concepts will help you be successful as you work through this unit.

Complete the problems on a separate sheet of paper.

1) What are all of the sets of numbers under the set of real numbers? Write out the subsets and/or draw a diagram.

2) What is a rational number?

3) Write the equation in terms of y. $5(3x + y) = p$

4) Kwame earns $200 per month at an after school job and has already saved $750. Dahlia has $75 saved and earns $350 each month. How many months will it take for Kwame and Dahlia to have the same amount of money? Write an equation and solve.

5) Solve. $|5x + 7| = -9$

6) Solve. $6\left|\frac{3}{4}p - 2\right| = 18$

7) The range of weights for a box of multigrain crackers can be plus or minus 0.75 ounces from the median weight of 14 ounces. Write an absolute value inequality to find the range of weights and then solve.

8) Write the symbol for the two inequalities that represent the graph.

    3 $-n$  -3

9) Explain why the inequalities in problem 8 are represented by the same graph.

10) What is the average quiz score for the afternoon class (to the nearest unit)?

11) What is the average quiz score for the morning class (to the nearest unit)?

12) Does it appear that the quiz questions favored one of the groups? Explain.

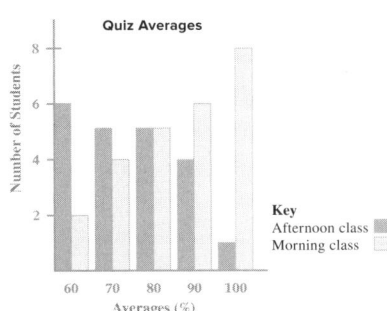

Multiple Choice

13) Which graph represents the solution for the given inequality $|3x + 4| > 7$?

   A)

   B)

   C)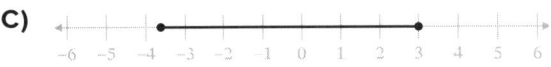

   D)

14) Solve.

   $-\frac{1}{3}\left(\frac{1}{2}x + 5\right) = -1$

   A) -4

   B) $-16, -4$

   C) no solution

   D) all real numbers

Lesson 6
Understanding Data

Outline

Part A
Data Calculations

- Measures of Center
- Measures of Spread
- Box Plots
- Dot Plots and Bar Graphs
- Histograms

Part B
Interpreting Data

- The Shape of Data Sets
- Standard Deviation
- Outliers
- Comparing Data Sets

Targeted Review

Vocabulary

- data
- quantitative data
- categorical data
- mean
- median
- mode
- range
- spread
- interquartile range (IQR)
- standard deviation (σ)
- five-number summary
- dot plot
- bar graph
- histogram
- box plot
- skewed data
- right-skewed
- left-skewed
- normal distribution
- bell curve
- deviation
- 68-95-99.7 Rule
- outliers

6A

Part A: Data Calculations

Objectives

In this part of the lesson, you will learn about data calculations.

By the end of this lesson you will be able to do the following:

- Match a data set to a graph, including dot plots, histograms, and box plots.
- Calculate measures of spread, including range, interquartile range, and the five-number summary.
- Calculate the measures of center: mean, median, and mode.

Why?

From figuring out which store has the best discount on your favorite sneakers to the average age of people that purchase your company's product, being able to understand and calculate data will impact your world, personally and professionally, throughout your life.

Warm Up

Use the table below to answer the following questions.

Last Week's High Temperatures (°F)

S	M	T	W	Th	F	Sa
85	80	60	75	55	45	55

1) What was the median temperature last week?

2) What was the mean temperature last week?

3) What was the mode temperature last week?

4) What was the range of temperatures last week?

EXPLORE 6A

▷ Measures of Center

- _____ is a collection, or set, of information that can be quantitative or categorical.

- _____ data is numerical.

- _____ data is data that is divided into categories or groups.

Measures of Center:

- _____ is the average.

- _____ is the middle element of the data set when the data set is ordered.

- _____ is the most frequently occurring element in a data set.

- When an element *larger* than the mean is added to a data set, the mean _____.

- When an element *smaller* than the mean is added to a data set, the mean _____.

- When an element that is the _____ as the mean is added to the data set, the mean remains the same.

Example 1

Find the average when elements are added to the data set.

After taking five tests, Jill had a mean test score of 85. What would her average be if she scored 75 on the next test? Round to the nearest unit.

$85 \cdot 5 =$ Total points earned on 5 tests   Setup with calculator:

$+ 75 =$ New point total $\bar{x} = \dfrac{(85 \cdot 5) + 75}{6}$

$\dfrac{}{6} =$ New average of 6 tests

> Remember to enter the entire problem into the calculator at one time.

The new average is about _____ .

6A EXPLORE

Example 2

Find the element that will result in the desired average for the data set.

The basketball team had an average score of 90 points between their first four games. They wanted their average score after five games to be 95 points. What would the final score in the next game need to be to achieve this average?

n: needed score for new average

$4 \cdot 90 = 360$	◂ Total points in four games
$5 \cdot 95 = 475$	◂ New average with 95-point average in five games

$n + 360 = 475$ ◂ Equation to find the next game's score

$-360 \quad -360$ ◂ Addition Property of Equality

$n = 115$

To reach a final average of 95 after five games, the team must score 115 points.

This problem can also be solved using an equation:

$$\frac{(4 \cdot 90) + n}{5} = 95$$

$$\frac{360 + n}{5} = \quad$$ ◂ Write 95 as a ratio

$(360 + n)(1) = (5)(95)$ ◂ Cross product, simplify terms

$360 + n =$ ◂ Addition Property of Equality

$n =$

> Either method to solve is correct. You should determine the strategy that makes the most sense to you.

☑ Checkpoint

Bo has a class average of 88 after taking four tests. Bo was hoping for an average of 90 for the grading period. What score does Bo need to earn on the next test to have a 90% average?

Pick the correct equation and use it to solve.

A) $\bar{x} = \frac{(88 \cdot 4) + 90}{5}$

B) $90 = \frac{(88 \cdot 4) + n}{5}$

EXPLORE 6A

▶ Measures of Spread

- The _____ of a data set represents how far apart the elements of the data set are from one another.

Measures of Spread:

- Range is found by subtracting the _____ value from the _____ value.

- A quartile is a _____, or 25% of the data.

    - The lower quartile, _____, represents the bottom 25% of the data set.

    - The upper quartile, _____, represents the top 25% of the data set.

- Interquartile range (IQR) represents the spread of the middle _____ of a data set and measures the spread based on the _____ value.

- IQR = _____ − _____

- _____ is the average distance of elements in the data set from the mean.

- The parts of the five-number summary are:

    -
    -
    -
    -

Example 3

Find the interquartile range for: {2, 3, 4, 4, 5, 5, 7, 7, 8, 10, 11, 11, 11, 12}

Plan Write the data set in order
Calculate the median, Q1, Q3
Calculate the IQR

Implement

{2, 3, 4, 4, 5, 5, 7, 7, 8, 10, 11, 11, 11, 12}

6A EXPLORE

Example 4

Find the interquartile range for: {19, 19, 20, 21, 21, 22, 23, 23, 24, 25, 28, 28, 29}

Plan Write the data set in order
Calculate the median, Q1, Q3
Calculate the IQR

Implement

{19, 19, 20, 21, 21, 22, 23, 23, 24, 25, 28, 28, 29}

☑ Checkpoint

Calculate the five-number summary, interquartile range, and range.
{28, 37, 39, 43, 46, 47, 22, 34, 38}

▶ Box Plots

- _____ divide data sets into four parts, each with

  the _____ of elements.

- The _____ is

  calculated to determine the construction of the box plot.

- In a box plot, the box represents the

  _____ of the data elements.

- In a box plot, _____ are drawn from the minimum to the first quartile and from

  the third quartile to the maximum.

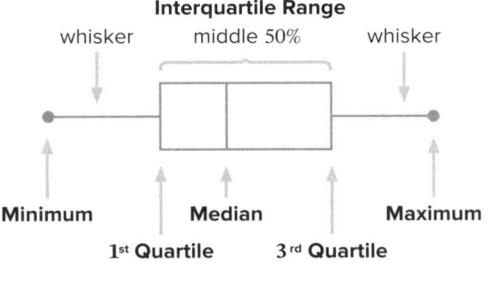

Interquartile Range
whisker — middle 50% — whisker
Minimum — 1st Quartile — Median — 3rd Quartile — Maximum

- The median is the middle number in the data set but is _____ in the middle of the box when graphed.

- Box plots are useful in showing the _____ of the data.

- _____ and _____ are measures of spread that can be calculated from a box plot.

- There are _____ sections in a box plot.

- Each section represents _____ of the data set and has the same number of elements in each part.

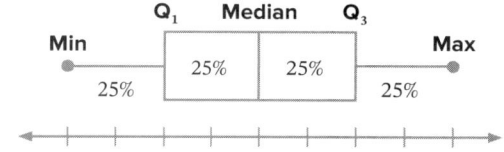

- If the _____ of elements in the data set is known, it can be divided into _____ parts, or quarters, to determine the number of elements in each section of the box plot.

Example 5

Mrs. Madigan asked 24 of her students to record the time they spent reading for homework last night.

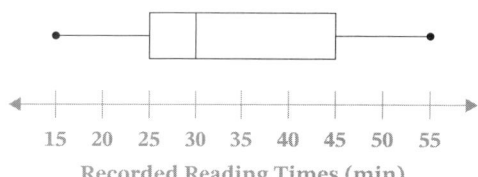

Recorded Reading Times (min)

Use the box plot to answer the questions.

A) What is the median reading time?

B) What is the interquartile range?

C) What percentage of students read between 25 and 45 minutes?

D) What percentage of students read for more than 25 minutes?

E) What quartile contains elements from 15 – 25 minutes?

F) How many students are in the upper quartile?

G) What is the average?

H) How many students read for less than 30 minutes?

I) Does the box plot represent measures of center or spread better? Explain.

6A EXPLORE

✓ Checkpoint

Use the box plot to find the range and interquartile range (IQR). Determine the average or explain why it cannot be calculated.

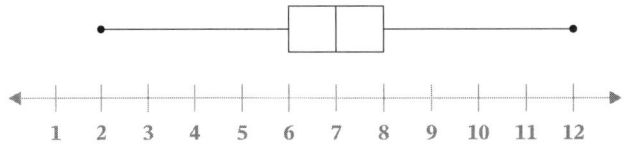

▶ Dot Plots and Bar Graphs

Dot Plots

- Determining which _____ to use is just as important as calculating measures of center and spread.

- Graphing data incorrectly can lead to _____.

- _____ use basic shapes to represent each element in a data set.

- Dot plots are useful for representing _____ data sets with _____ elements.

Bar Graphs

- _____ are used to compare categories of data and usually have a vertical axis number line.

- Unlike dot plots, bar graphs can represent _____ counted values.

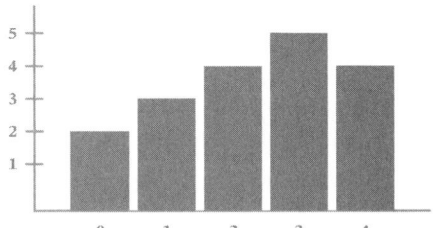

Example 6

Find the measures of center from the bar graph.
The bar graph recorded customer reviews at Harvey's Restaurant.

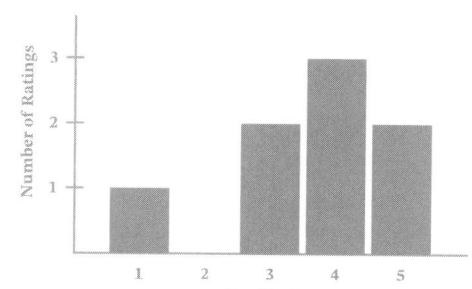

Plan   List the elements
Determine the mean, median, mode

Implement

Elements of the bar graph: {1, 3, 3, 4, 4, 4, 5, 5}

Mode:
The tallest bar and the number that repeats the most is the mode.

Median:

Find the middle number when the data set is listed in order:

{1, 3, 3, 4, 4, 4, 5, 5}

Mean:

Add all of the values and divide by the total number of elements.

$$\frac{(1 \cdot 1) + (3 \cdot 2) + (4 \cdot 3) + (5 \cdot 2)}{8} =$$

☑ Checkpoint

Use the dot plot to determine the measures of center. Round to the nearest tenth where needed.

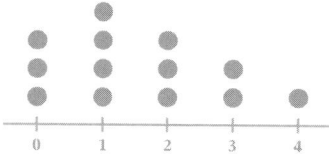

6A EXPLORE

▶ Histograms

- In a _____, the horizontal axis has equal intervals of data instead of categories.

- _____ on a histogram meet but never overlap.

- In a histogram, each element of the _____, will fit into only one interval.

- The _____ axis of a histogram identifies the number of elements in each interval.

- Histograms can be used to _____ the intervals created for the data set, but cannot be used to find the exact _____ or _____ for the data set.

- Without being given the actual data set, values calculated from a histogram will be an _____ only.

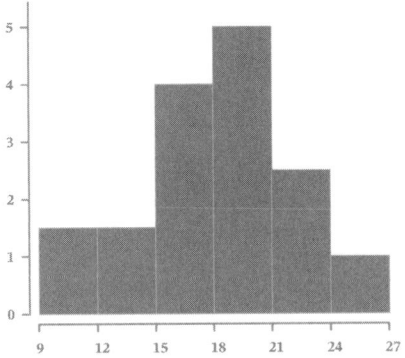

Example 7

Use the histogram to determine the estimated medan and range.

On field day, the students recorded their times in minutes to complete the obstacle course. They used the data to create a histogram.

Estimated range:

largest value − smallest value

Estimated median:

total number of elements: 20

middle elements: 10th and 11th (2 intervals from left and 3 intervals from right)

estimated element values:

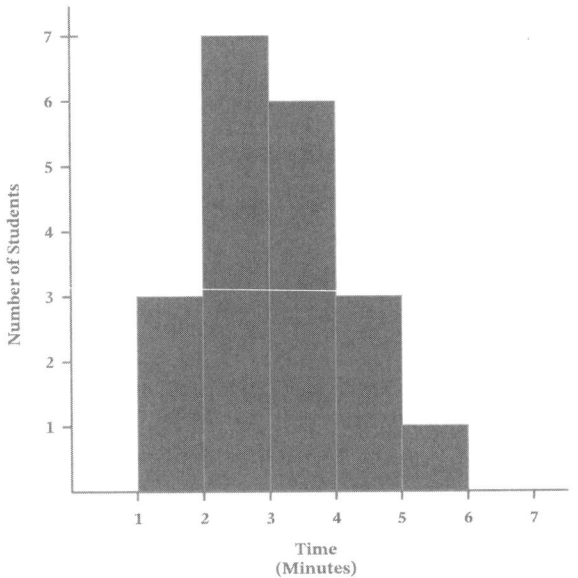

☑ Checkpoint

Use Example 7 to find the estimated mean from the histogram. Explain why the mean is an estimate.

6A PRACTICE 1

Practice 1

Complete the problems on a separate sheet of paper. Round to the nearest hundredth where needed.

1) Seven students went on a field trip. Their teacher suggested they take extra money to spend in the gift shop. The set below represents how much extra money each student brought.

   Using only the dot plot determine mean, median, and mode.

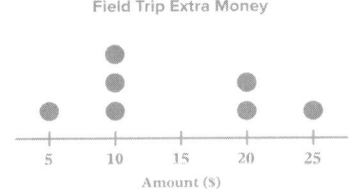

Mrs. McCready analyzed the test she had given her students using a box plot. The scores for the test were:

{60, 90, 80, 90, 70, 65, 100, 55}.

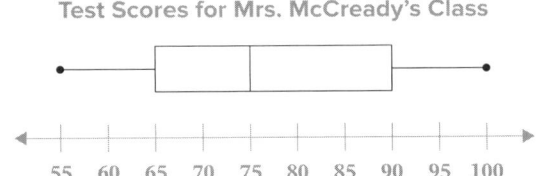

2) Determine the average and the median.

3) Mrs. McCready considers a passing score to be 75%. What percentage of her class passed based on this determination? How many students passed the test?

Determine how the value of the new element will affect the mean (increase, decrease, remain the same). Then find the new average of the data set after adding the new elements.

4) Original data set

   Number of elements: 3
   Mean: 52
   New element: 52

5) Original data set

   Number of elements: 24
   Mean: 28.3
   New elements: 6.4, 4.3

Find the value of the element necessary to arrive at the desired mean.

6) Original data set

   Number of elements: 8
   Original Average: 27.5
   Desired Average: 30

7) Original data set

   Number of elements: 12
   Original Average: 90
   Desired Average: 90

Determine the graph that best matches the given situation.

8) Benjamin is applying for a scholarship. To qualify, he needs to score in the top 50% of applicants on a specific test. Which type of graph is the best for Benjamin to see the top 50% of scores for the test?

9) Danielle was researching the cyclists that traveled down the hill in front of her home. She found that cyclists traveled down the hill at various speeds ranging from 3.2 miles per hour (mph) to 25.7 mph. Which type of graph would best represent the number of cyclists traveling down the hill at speeds within 5 mph intervals?

Match the data set to a graph.

A) 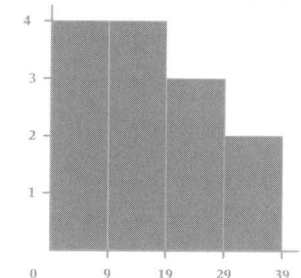 B) 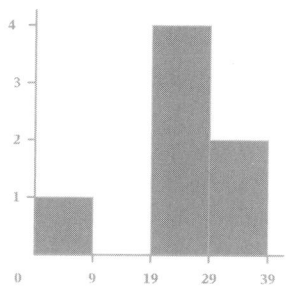 C)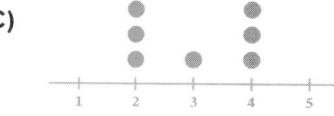

_____ 10) {20, 32, 25, 35, 9, 22, 25}

_____ 11) {15, 21, 26, 25, 7, 6, 10, 17, 10, 30, 8, 35, 3}

_____ 12) {2, 3, 4, 2, 2, 4, 4}

Use the box plot to answer problems 13-16. It represents the commutes of 13 people at a local office.

13) What percentage of people have a commute with a distance greater than 15 miles?

14) What percentage of people commute between 8 and 25 miles to work?

15) How many people are represented in the first quartile?

16) What is the IQR?

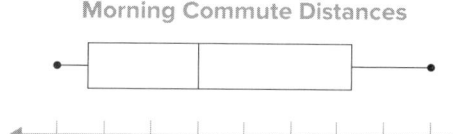

Morning Commute Distances
Distance (miles)

17) Mary-Anne listed her age and the age of her four siblings: {4, 7, 7, 13, 15}
In six years, will the following measures increase, decrease, or remain the same?

mean   range
median   IQR
mode

Name the measure of center or spread that best matches the scenario.

18) Anderson said on his farm that most cows have one calf each time.

19) Elizabeth found the difference between the highest and lowest temperature in March was 30 degrees.

20) Dodge has an overall grade in Algebra of 92%

21) What are the *similarities and differences* between the mean and median of a data set?

6A MASTERY CHECK

Mastery Check

✎ Show What You Know

Mr. Stephens recorded the scores of the last math quiz for his class. A total of 25 points could be earned for the quiz.

Class quiz scores: {15, 18, 19, 19, 19, 19, 21, 24, 24, 24, 25, 25}

A) Calculate the five-number summary for the data set.

B) Create a box plot for the class quiz scores.

C) What percent of the class earned grades higher than 75%? Explain.

D) Write an absolute value inequality that will represent all of the class quiz scores. Explain what the numbers in your inequality represent.

·ıl|ı· Say What You Know

In your own words, talk about what you have learned using the objectives for this part of the lesson and your work on this page.

Practice 2

Complete the problems on a separate sheet of paper. Round to the nearest hundredth where needed.

1) How can you determine the mean and median from a bar graph or dot plot?

2) Determine the measures of center for the bar graph.

3) Can you determine the interquartile range from the given bar graph? Explain.

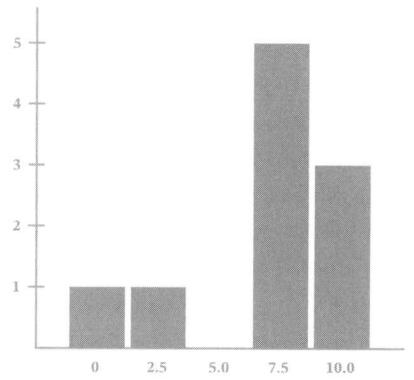

Determine how the value of the new element will affect the mean (increase, decrease, remain the same). Then find the new average of the data set after adding the new elements.

4) Original data set

   Number of elements: 8
   Mean: 25
   New elements: 48 and 22

5) Original data set

   Number of elements: 10
   Mean 12.5
   New elements: 15, 22, 12, 16, 20

Find the value of the element necessary to arrive at the desired mean.

6) Original data set

   Number of elements: 4
   Original Average: 89
   Desired Average: 90

7) Original data set

   Number of elements: 20
   Original Average: 89
   Desired Average: 90

Determine the graph that will best represent the given situation.

8) Mrs.Collins wants to compare the data she collected from forty-five third graders on their favorite colors. The categories are blue, red, green, yellow.

9) Dr. Jones collected data about the heights of old-growth trees. He needs to analyze the data once it is broken into quarters.

6A PRACTICE 2

Match the graph to its data set.

_____ 10) 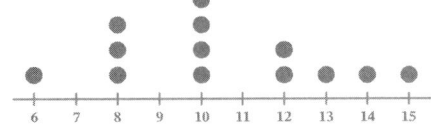   A) {6, 8, 11, 12, 13, 14, 15, 12, 11, 11, 8, 8, 11}

_____ 11) 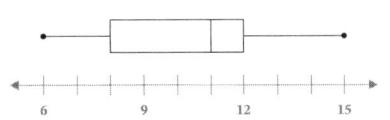   B) {6, 8, 9, 12, 13, 14, 15, 12, 9, 14, 8, 8, 10, 8}

_____ 12) 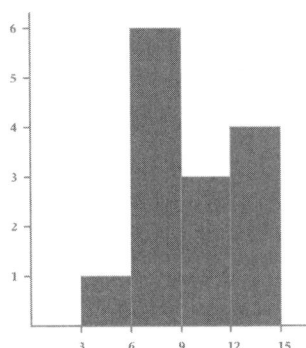   C) {6, 8, 10, 12, 13, 14, 15, 12, 10, 10, 8, 8, 10}

Mr. Cotes's ice hockey team has currently played 48 games. The box plot represents the numbers of goals scored at each of the 48 games.

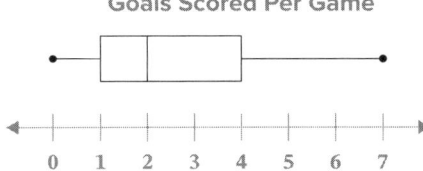

13) What is the range in goals that represents the IQR?

14) What percentage of games had between 0 and 1 goals scored?

15) How many games had goals over 4 goals scored?

16) Mr. Cotes told the team he used the box plot to calculate the average goals per game. Explain why this is not the correct way to calculate the average.

---

17) Centimeters of rainfall were recorded for the same five weeks in 2012 and 2019.

   2012: {1, 1.5, 2, 0, 0.5}   2019: {0, 0.5, 0, 1, 1.5}

   In 2019, did the following measures increase, decrease, or remain the same?

   mean   range
   median   IQR

Name the measure of center or spread that best describes the scenario.

18) The middle half of the class's scores had a range of 5.

19) The most common grade in the class was 88%.

20) The middle score for the class was 87%.

21) What added value(s) will leave the mean unchanged? Explain.

Part B: Interpreting Data

Objectives

In this part of the lesson, you will learn about interpreting data.

By the end of this lesson, you will be able to do the following:

- ✓ Use the 68-95-99.7 Rule and bell curves to analyze standard deviations.
- ✓ Use the outlier formula to determine if a data set contains outliers.
- ✓ Compare center and spread for multiple data sets, including their outliers.

Why?

Understanding concepts like outliers and standard deviation will allow you to better understand and use the data you encounter in algebra and your everyday life.

Warm Up

Answer the following questions using the data set: {28, 37, 39, 43, 46, 47}.

mean = 40   median = 41   range = 19

1) If 40 was added to the set, would the mean increase, decrease, or stay the same?

2) If 40 were added to the set, would the median increase, decrease, or stay the same?

3) If 68 was added to the original set, compared to the original mean, would the mean increase, decrease, or stay the same?

4) If 68 were added to the original set, compared to the original range, would the new range increase, decrease, or stay the same?

The Shape of Data Sets

- The _____ of a data set reveals the visual patterns in the data.

- _____ data has more points on one side than the other.

6B EXPLORE

- The data in the graphs below is _____-skewed.

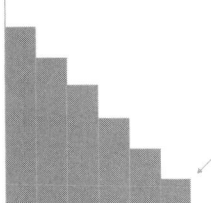

- The data in the graphs below is _____-skewed.

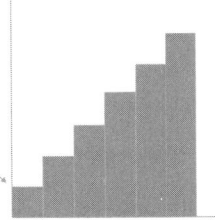

- Data that is normally distributed makes a _____ and has no skew.
  - This is also called a _____ distribution.

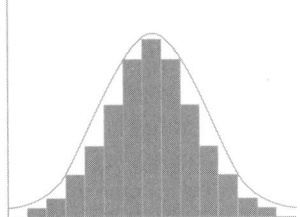

✓ Checkpoint

Make a quick sketch of each of the following types of graphs as histograms.

Left skewed   Normal   Right skewed

EXPLORE 6B

▶ Standard Deviation

- _____ is a measure of spread based on the mean of a data set.

- Standard deviation represents the average deviation, or _____, of elements from the _____.

- Because standard deviation is a measure of distance, it is _____ positive.

- The probability of occurrences within normally distributed data and bell curves follows a pattern called the _____ Rule.

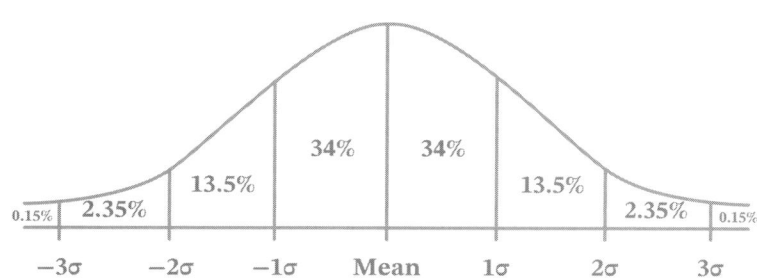

- In the 68-95-99.7 Rule:

    - values that are within _____ standard deviation(s) from the mean will make up 68% of the data.

    - values that are within _____ standard deviation(s) from the mean will make up 95% of the data.

    - values that are within _____ standard deviation(s) from the mean will make up 99.7% of the data.

- Each _____ line on a bell curve represents a standard deviation above or below the mean.

6B EXPLORE

Example 1

Fill in the numbers to represent one, two, and three standard deviations from the mean. Then use the normal distribution to answer the following questions.

The mean of the data set for the sum of ten dice is 35 and the standard deviation is approximately 5.

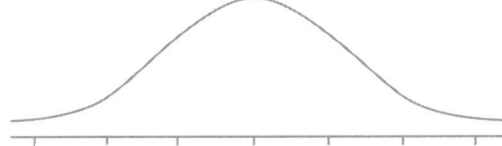

A) What percent chance is there that the sum will be one standard deviation from the mean?

B) What is the range within two standard deviations? What percentage of the data does this represent?

Two standard deviations represent _____ % of the data.

C) What is the percent chance that the sum of all ten dice will be 60?

D) What is the percent chance that the sum of all ten dice will be 40 or less?

$30 < x < 40 =$

$25 < x < 30 =$

$20 < x < 25 =$

$x < 20 =$

> These percentages will remain true for all normally distributed data.

☑ Checkpoint

Use the normal distribution to answer the following questions.
The normal distribution represents the average height of a candy floss tree in feet.

A) What is the average height of the tree?

B) What is the standard deviation?

C) If you see a candy floss tree in the woods, what chance is there that the tree will be between 8 and 10 feet tall?

D) What is the name of the rule for normally distributed graphs?

EXPLORE 6B

▶ Outliers

- _____ are data elements that are outside the overall pattern of the data set.

- The _____ is used to back up your instinct mathematically.

Outlier Formula:

lower outlier $< Q1 - (1.5 \cdot IQR)$

upper outlier $> Q3 + (1.5 \cdot IQR)$

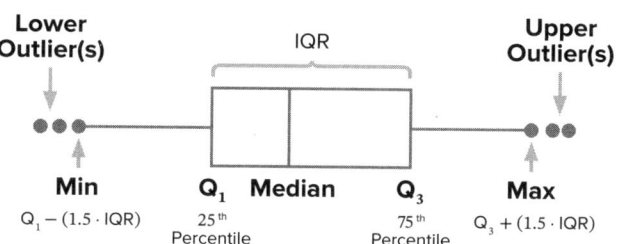

- To solve for outliers using the formula the _____ must be calculated.

- When outliers are not calculated, they can cause the data to be skewed unintentionally.

Example 2

Mr. Tyrone wants to know if there are any outliers present in the last test given to his students. He believes that the lowest and the highest scores are both outliers. He grades the number of points students earned out of 100.

Grades: {55, 73, 75, 78, 80, 83, 85, 85, 88, 89, 90, 92, 96, 99, 100}

A) Use the outlier formula to help determine if any outliers are present in the data.

Five-Number Summary
- Min: 55
- Q1: 78
- Med: 85
- Q3: 92
- Max: 100

IQR: $92 - 78 = 14$

Lower Outlier $< Q1 - (1.5 \cdot IQR)$

Lower Outlier $<$

Lower Outlier $<$

The student that earned a 55 is an outlier.

Upper Outlier $> Q3 + (1.5 \cdot IQR)$

Upper Outlier $>$

Upper Outlier $>$

There are no upper outliers.

Mr. Tyrone wants to know what effect the outlier has on the class average. He calculated the mean before determining the outliers and found it was about 84.53.

B) Calculate the average without the outlier.

$$\frac{73 + 75 + 78 + 80 + 83 + 85 + 85 + 88 + 89 + 90 + 92 + 96 + 99 + 100}{14} =$$

C) How does removing the outlier affect the average?

6B EXPLORE

☑ Checkpoint

Determine the lower and upper bounds for the outliers using the data set. Are there any outliers in the data set? If so, name the outlier(s).

$$\{16, 18, 19, 20, 20, 21, 32\}$$

▶ Comparing Data Sets

■ Some examples of how comparing data sets can be useful in everyday life are:

_____ , _____ ,

and _____ .

Example 3

Use the graph to compare data sets.

Mrs. Meyer's Algebra class and Mrs. Wolfe's Algebra class were debating about which class did better on the last test. Not wanting to share individual test scores, the teachers made a stacked box plot for students to use to back up their arguments mathematically.

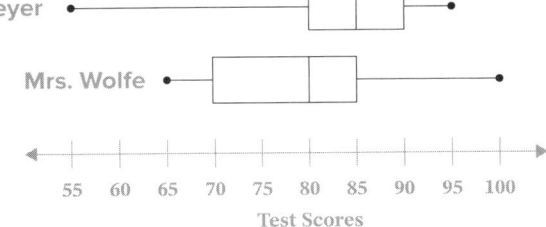

Implement

Mrs. Meyer	Mrs. Wolfe
Min:	Min:
Q1	Q1
Med:	Med:
Q3:	Q3:
Max:	Max:
Range:	Range:
IQR:	IQR:

Explain

116 Lesson 6: Understanding Data > Part B: Interpreting Data > Explore

☑ Checkpoint

Use the measures of center to determine which class grew more according to the data. Which measure best indicates this?

Two elementary classrooms measured their heights in centimeters at the start and end of the school year. Since the growth range was the same for both classes, the teachers decided to use measures of center to determine which class had a more significant height change.

Growth in Centimeters

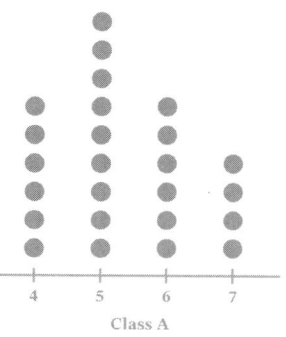
Class A

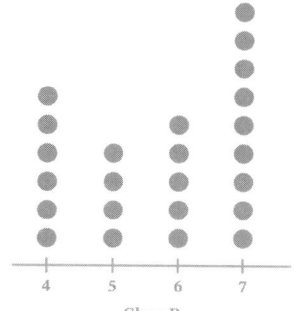
Class B

mean = 4.84
median = 5 cm
mode = 5 cm

mean = 5.64
median = 6 cm
mode = 7 cm

6B · PRACTICE 1

✏️ Practice 1

Complete the problems on a separate sheet of paper.

Label the shape of the graph as left-skewed, right-skewed, or normally distributed.

1)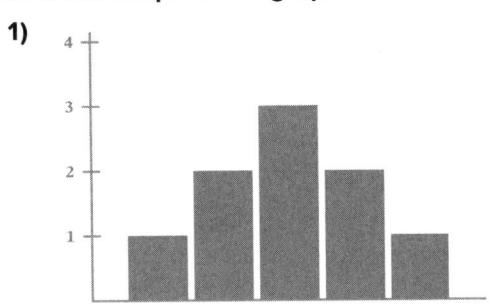

2)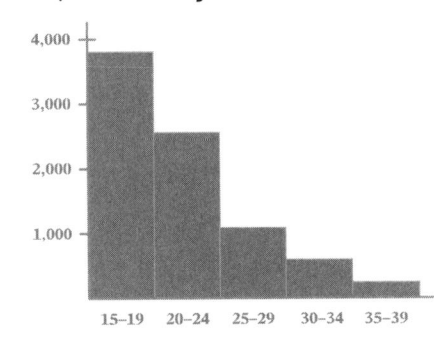

3) What is the 68-95-99.7 Rule?

Use the graph for problems 4–6.

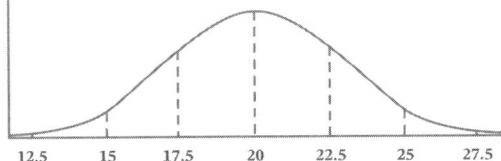

4) What is the mean of this data set?

5) What is the standard deviation of this data set?

6) Write the inequality that would represent about 68% of the elements in the data set.

7) Draw a normal curve. Then label the graph for the data set when the mean = 22, and the standard deviation = 3.5.

Use the graph for problems 8–11.

It takes an average time of 65 days for a certain flowering plant to grow and bloom.

8) About what percentage of the plants bloom between day 55 and 75?

9) About what percentage of plants bloom before day 45?

10) What percent chance would a plant bloom after day 55?

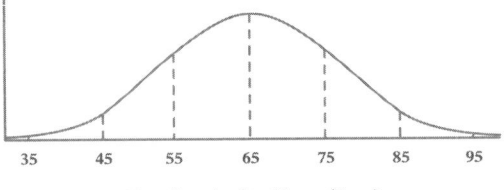

Time Required to Bloom (Days)

11) Mr. Miller's class hypothesized that the plant would bloom between day 35 and day 55. Ms. Nance's class hypothesized that the plant would bloom between day 55 and day 75. The range of days is equal for both classes. What class has a better chance of having a correct hypothesis? Explain.

12) The data set represents Allison's test scores for the last grading period in numerical order: {60, 82, 85, 87, 90, 91, 92, 96}. Determine the lower and upper outliers for the data set. Name the outliers if any occur.

Mikey recorded the number of birds that landed on a bird feeder for ten days during the same hour each day. Then he listed his data set from greatest to least:

$$\{16, 10, 9, 8, 6, 5, 5, 3, 2, 2\}$$

13) Mikey believes that 16 is an outlier because it seems far above the other numbers. Determine if this is correct using the outlier formula.

14) On the eleventh day, Mikey saw 22 birds on the bird feeder. If he included this number in his data set, what would happen to the mean?

In golf, a lower score is better than a high score, and par is a score of 0. Rory and Fran each played sixteen rounds of golf and recorded their scores.

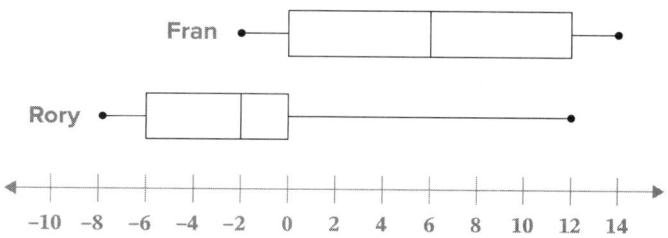

15) Find the five-number summary as well as the range and interquartile range for each golfer.

16) Fran claims that he is a better golfer because his range in scores is lower. Based on the box plots, who is the better golfer?

Two coffee shops were competing to be the best cup of coffee in the town. The plan was to give the win to the shop that earned the most votes but both received the same number of reviews and had the same mode.

17) Calculate the mean and median for both coffee shops.

18) Use the calculations from the previous question to determine the better coffee shop.

6B MASTERY CHECK

Mastery Check

Show What You Know

Two groups of students were asked to record the time spent on homework last week.

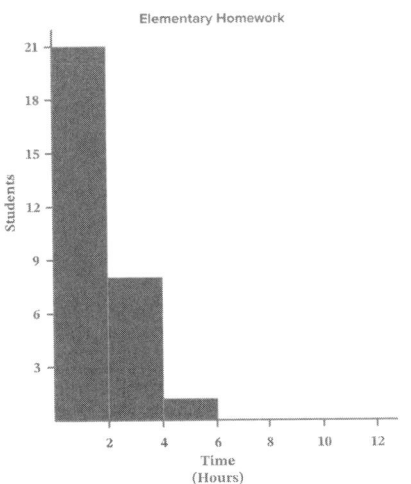
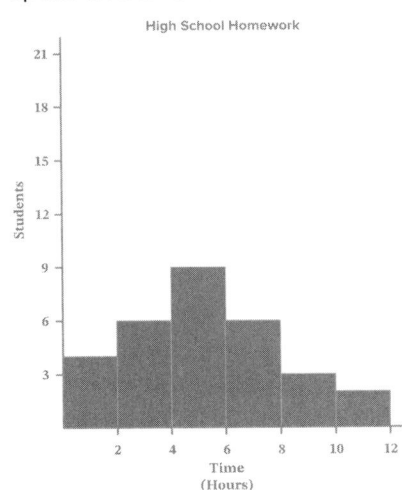

A) Without doing any calculations, where would you expect the measures of center to be for the *elementary* students? Is the data skewed or normal in its distribution? Explain.

B) In general, how much homework do the elementary students have each week?

C) Without doing any calculations, where would you expect the measures of center to be for the *high school* students? Is the data skewed or normal in its distribution? Explain.

D) Suppose the high school teachers surveyed 1,000 students. The average time for this group was 5 hours and the standard deviation was 2. Label the normal curve correctly. What is the range of hours that will include 95% of the students surveyed?

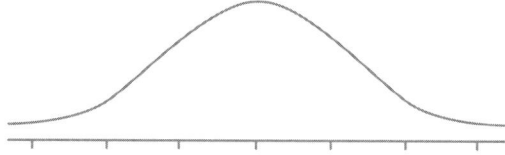

E) Three students say this cannot be right because they spend 16 hours a week doing homework. How would you identify this group of students?

⁙⁙ Say What You Know

In your own words, talk about what you have learned using the objectives for this part of the lesson and your work on this page.

6B PRACTICE 2

✏ Practice 2

Complete the problems on a separate sheet of paper.

Label the shape of the graph as left-skewed, right-skewed, or normally distributed.

1)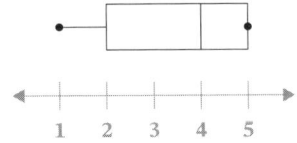

2)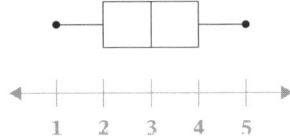

3) According to the 68-95-99.7 Rule, what percentage of values will fall within two standard deviations from the mean? What percentage will fall within one standard deviation from the mean?

Use the graph for problems 4–6.

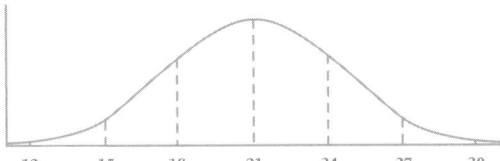

4) What is the mean of this data set?

5) What is the standard deviation of this data set?

6) Write the inequality that would represent about 68% of the elements in the data set.

7) Draw a normal curve. Then label the graph for the data set when the mean = 63, and the standard deviation = 5

Use the graph for problems 8–11.

Students in primary grades were asked to read for 20 minutes on average each night.

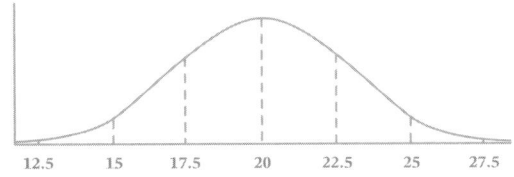

8) About what percentage of students read less than 25 minutes?

9) About what percentage of students read between 15 and 22.5 minutes?

10) About what percentage of students read for more than 27.5 minutes each night?

11) Ms. Casey asked her class to read for 25 minutes or more every night. What is the likelihood that all of her students will read 25 minutes or more? Would there be a better range of times to get more students to read? Explain.

The soccer team statistician listed the number of goals scored this season:

$$\{0, 0, 2, 3, 3, 3, 4, 4, 5, 6\}$$

When calculating the average, the statistician thought the number seemed lower than expected given the number of goals scored during the season.

12) Determine the lower and upper outliers for the data set. Name the outliers if any occur.

13) What is the cause for the average being lower than expected?

Mr. Riley's biology class asked for a retest. Mr. Riley said that a retest was unnecessary based on the class average of 75.5 that he calculated from the following test scores:

{52, 58, 60, 62, 62, 63, 65, 65, 68, 100, 100}

14) Use what you know about outliers to convince Mr. Riley that the average is not a good indication of how the class performed.

15) Mr. Riley decides to remove the outlier before recalculating the average. What will this do to the average? Should he allow a retest based on the new average? Explain.

There were 24 Eagles players and 36 Tigers players surveyed on the two volleyball teams. Compare the data by answering the following questions.

16) How many players on the Tigers have a height of 68 inches or less?

17) What percentage of Eagles players are 68 inches or shorter?

18) How many Eagles players are in each 25% of the data? How many Tigers players are in each 25% of the data?

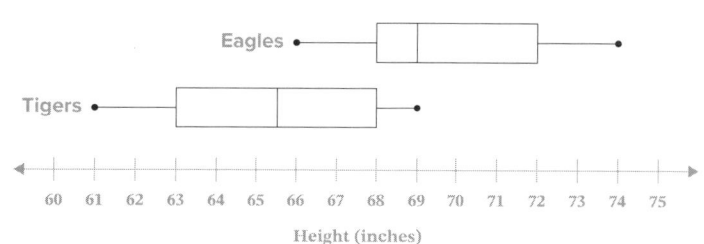

19) Explain how the data shows that Eagles players are overall taller than Tigers players.

A group of 600 people was asked to determine how long it takes them to get ready for school in minutes each morning. Three hundred surveyed were high school students, and the rest were college students.

20) Explain why using a dot plot would not be the best way to represent the data.

21) Use the data in the table to explain how the two groups can have different means but the same median and mode.

High School Student	College Student
mean = 37 min	mean = 29 min
median = 30 min	median = 30 min
mode = 30 min	mode = 30 min

TARGETED REVIEW 6

◎ Targeted Review

In the Targeted Review, you will practice topics you have mastered in earlier lessons. Reviewing these concepts will help you be successful as you work through this unit.

Complete the problems on a separate sheet of paper.

1) Write all numbers as integers. Solve.
$$\tfrac{2}{5}x = \tfrac{1}{3}(x + 6)$$

2) Name all of the sets of numbers to which your solution to problem 1 belongs using math shorthand.

Solve. Justify each step with algebraic properties.

3) $3 + \left|\tfrac{1}{2}n - 8\right| = 5$

4) $-4|3x + 5| - 6 = 1$

Solve. Graph your solution on a number line.

5) $\tfrac{4}{5} < 2x - 7 \leq 5$

6) $-\left|\tfrac{2}{3}x + 1\right| < -5$

Solve.

7) $\dfrac{3x + 4}{5} = \dfrac{x}{8}$

8) $\dfrac{5}{3} = \dfrac{x - 6}{4}$

Multiple Choice

____ 9) Write in terms of x.
$$2\left(5y - \tfrac{1}{3}x\right) = C$$

    A) $x = -\tfrac{C}{6} - \tfrac{5}{3}y$

    B) $x = -\tfrac{C}{6} + \tfrac{5}{3}y$

    C) $x = -\tfrac{3C}{2} - 15y$

    D) $x = -\tfrac{3C}{2} + 15y$

____ 10) The goal of the cross country team was for everyone to finish the race (r) in 44 minutes. The entire team finished within 7 minutes of this time. Choose the absolute value equation that represents the minimum and maximum times for the race.

    A) $|r + 7| = 44$

    B) $|r - 7| = 44$

    C) $|r + 44| = 7$

    D) $|r - 44| = 7$

____ 11) Solve:
$4|x - 2| + 1 \leq 9$

    A) $x \leq 4$

    B) $x \geq 0$

    C) $0 \leq x \leq 4$

    D) $x \leq 0$ OR $x \geq 4$

____ 12) How fast is an object moving in miles per hour when it is traveling 60 feet per second? Round to the nearest whole number.

    A) 40 mph

    B) 41 mph

    C) 60 mph

    D) 720 mph

UNIT 2 Record Keeping   Name: Algebra 1

Lesson		Part	Guided Notes	Practice 1	Practice 2	Mastery Check	Targeted Review	Lesson Test
7 Functions	A	Relations and Functions						
	B	Understanding Functions						
8 Using Graphs	A	Intercepts and Slope from a Graph						
	B	Translating the Linear Parent Function						
9 Slope and Linear Functions	A	Slope and Graphed Scenarios						
	B	Point-Slope Form and Slope-Intercept Form						
10 Writing Linear Functions	A	Writing Equations in Slope-Intercept Form						
	B	Applications of Linear Equations						

Lesson Objectives

Lesson 7: Part A
- ☐ Find the domain and range of a relation from a graph, table, and mapping.
- ☐ Express the definition of a function using words and diagrams.

Lesson 7: Part B
- ☐ Write equations with two variables in function notation.
- ☐ Identify variables as dependent or independent.
- ☐ Evaluate a function for the dependent variable given a set of values for the independent variable.
- ☐ Determine whether a specific point is a solution for a function presented as a graph, table, or equation.

Lesson 8: Part A
- ☐ Identify intercepts of a line from a graph, table, or equation.
- ☐ Given a graph of a line, determine the slope using slope triangles or rise over run.
- ☐ Plot the graph of a line given a point and the slope.

Lesson 8: Part B
- ☐ Express the linear parent function in all forms (table, graph, and equation).
- ☐ Demonstrate translations by a factor of b to a linear function.
- ☐ Explain how b translates a linear function up or down.

Lesson 9: Part A
- ☐ Use the slope formula, $m = \frac{\Delta y}{\Delta x}$, to calculate the slope of a line when given two points on the line.
- ☐ Describe a graph as a scenario using mathematical vocabulary and sketch a graph from a written scenario.

Lesson 9: Part B
- ☐ Write linear equations in point-slope form from a given graph or a point and the slope.
- ☐ Write linear equations in slope-intercept form from a graph or given the slope and the y-intercept.
- ☐ Graph equations on the coordinate plane in point-slope or slope-intercept form.

Lesson 10: Part A
- ☐ Write an equation in slope-intercept form given the slope and one point.
- ☐ Write an equation in slope-intercept form given two points.

Lesson 10: Part B
- ☐ Write an equation in slope-intercept form given any type of scenario.
- ☐ Explain what a given point, the slope, and the x- and y-intercept represent within the context of a word problem.

UNIT 2 Record Keeping   Name: Algebra 1

Lesson		Part		Guided Notes	Practice 1	Practice 2	Mastery Check	Targeted Review	Lesson Test
11	More Forms of Lines	A	Standard Form						
		B	Horizontal and Vertical Lines						
12	Parallel and Perpendicular Lines	A	Parallel Lines						
		B	Perpendicular Lines						
13	Scatter Plots	A	Correlation and Scatter Plots						
		B	The Line of Best Fit						
14	Types of Functions and Arithmetic Sequences	A	Continuous or Discrete Functions						
		B	Arithmetic Sequences						

Unit Test II   Date ____   Score ____

Lesson Objectives

Lesson 11: Part A
- ☐ Solve for the x- and y-intercepts from standard form and use the intercepts to create a graph of the line.
- ☐ Convert the equation of a line to standard form. Determine the slope and intercept formulas found in a linear equation in standard form.

Lesson 11: Part B
- ☐ Graph horizontal and vertical lines.
- ☐ Find the domain, range, slope, and y-intercept for horizontal and vertical lines.
- ☐ Determine the equation of horizontal and vertical lines that pass through a given point.

Lesson 12: Part A
- ☐ Identify parallel lines.
- ☐ Write the equation of a line that is parallel to another known line and passes through a given point.

Lesson 12: Part B
- ☐ Identify perpendicular lines.
- ☐ Write the equation of a line that is perpendicular to another known line and passes through a given point.

Lesson 13: Part A
- ☐ Identify a correlation of a scatter plot as strong or weak, positive or negative, or no correlation.
- ☐ Explain the meanings of correlations in real-life examples.
- ☐ Create a scatter plot with accurate scale, labels, and ordered pairs.

Lesson 13: Part B
- ☐ Estimate and draw the line of best fit for a set of data.
- ☐ Write the equation for the line of best fit in the requested form.
- ☐ Use the line of best fit to interpolate, extrapolate, and explain the context of the data.

Lesson 14: Part A
- ☐ Decide if a function is discrete or continuous and explain the difference.
- ☐ Use interval notation to define the domain and range of functions.
- ☐ Choose the most appropriate form of the equation of a line for a given scenario.

Lesson 14: Part B
- ☐ Describe the arithmetic sequence of a given set.
- ☐ Use a sequence to find additional terms.

Lesson 7
Functions

Outline

Part A
Relations and Functions

- The Coordinate Plane
- Relations
- Functions
- The Vertical Line Test

Part B
Understanding Functions

- Function Notation
- Independent and Dependent Variables
- Using a Function Rule

Targeted Review

Vocabulary

- coordinate plane
- origin
- quadrant
- coordinates
- ordered pairs
- relation
- domain
- range
- function
- mapping
- vertical line test
- independent variable
- dependent variable
- repeated substitution

7A EXPLORE

Part A: Relations and Functions

Objectives

In this part of the lesson, you will learn about relations and functions.

By the end of this lesson, you will be able to do the following:

- Find the domain and range of a relation from a graph, table, and mapping.
- Express the definition of a function using words and diagrams.

Why?

Seeing relations and functions in different ways will help you recognize them in word problems and in your world.

Warm Up

1) Plot the ordered pairs on the graph provided and label them with the letter indicated.

   A: $(-1, 2)$

   B: $(4, 0)$

   C: $(0, -4)$

   D: $(-2, -3)$

   E: $(0, 0)$

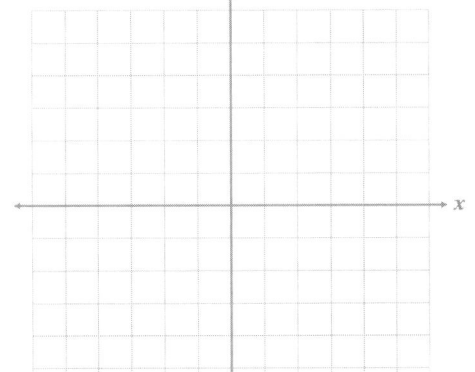

2) Which points align horizontally? Explain.

3) Which points align vertically? Explain.

▶ The Coordinate Plane

- The _____ is a two-dimensional graph formed by the x-axis and y-axis.

- The point $(0, 0)$ where the x-axis and y-axis intersect is called the _____.

- A quarter of the coordinate plane is called a _____.

- Quadrants are labeled _____ starting at the upper-right quadrant.

EXPLORE 7A

- The signs for an ordered pair in each quadrant are:

- _____ are sets of numbers that represent points on the coordinate plane.

- They are written as _____, where

  x represents the _____ direction of the point and y represents the

  _____ direction of the point.

  - x is positive in quadrants: _____

  - x is negative in quadrants: _____

  - y is positive in quadrants: _____

  - y is negative in quadrants: _____

- In this course, if the graph does not have a scale listed, it will be assumed the scale is _____.

Example 1

Create a rectangle on the coordinate plane using the three vertices provided. Plot point S to finish the rectangle.

Plan Visually determine where the fourth point belongs. Then, plot the point.

OR

Take the x-value from point T and the y-value from point R to make the new ordered pair for S. Then, plot the point.

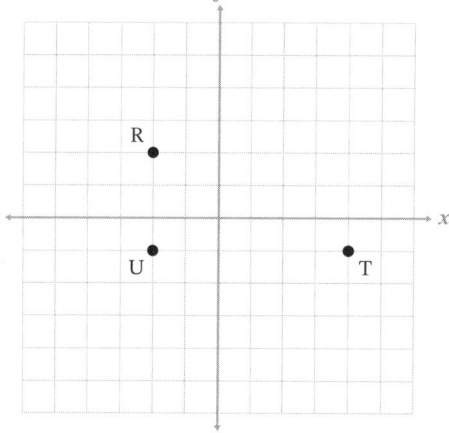

Point S:

Name the horizontal sides of the rectangle.

Name the vertical sides of the rectangle.

> Point is another name for an ordered pair, which are always ordered (x, y).

7A EXPLORE

> ☑ **Checkpoint**
>
> **Create a rectangle on the coordinate plane below using the three vertices provided. Plot the missing point to finish the rectangle.**
>
> Rectangle $MNOP$: $N(0, 4)$, $O(0, 0)$, $P(-5, 0)$
>
> Point M:
>
>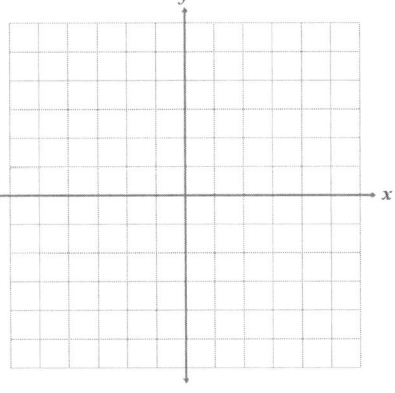
>
> Name the two horizontal line segments.

▶ Relations

- A _____ is a statement that represents a relationship between two variables and can be represented on a coordinate plane or written as a set of ordered pairs.

- The _____ of any relation is the set of possible x-coordinates (x, y).

- The _____ of any relation is the set of possible y-coordinates (x, y).

Example 2

Identify the domain and range for the given relation R.

$R: \{(3, 4), (5, 2), (-2, 1), (4, 6), (3, -5)\}$

Domain: $\{-2, 3, 4, 5\}$

Range: $\{-5, 1, 2, 4, 6\}$

The elements in the domain and range sets are written in order from _____.

Elements are not repeated in a set even if they appear more than once. The element 3 is used twice but only appears once in the domain.

EXPLORE 7A

Example 3

Given the graph of the relation, list the set of ordered pairs that make up the relation (Q). Then provide the domain and range of the relation.

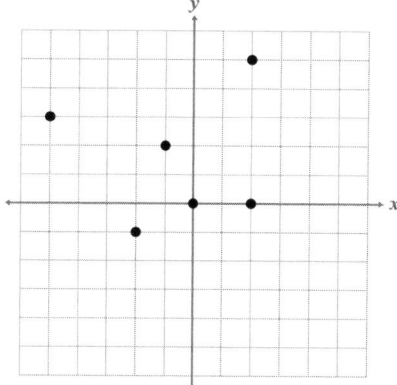

Domain:

Range:

☑ Checkpoint

Name the domain and range for the given relation.

$P: \{(-12, 8), (-8, 8), (-8, -8), (4, 2), (4, -8), (8, 8)\}$

Domain: Range:

▶ Functions

- A function is a particular type of relation that has _____ output for every input value.

- Every function is a relation, but _____ relation is a function.

- Since a function has one output for every input, the _____ values *cannot* repeat.

- Since a function has one output for every input, the _____ values can repeat.

- A _____ provides a visual representation of a relation.

- Creating a mapping:

    1) Draw and label the bubbles for the domain and range.

    2) Order the *x*-values.

    3) Draw arrows from the domain

7A EXPLORE

Example 4

Use the table to create a mapping. Determine if the relation is a function. Explain.

x	y
−2	4
−1	5
0	5
1	6

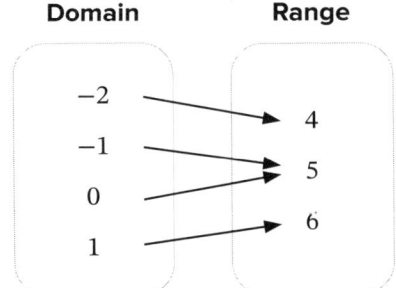

The relation is _____ because all of the domain values are _____.

Example 5

Determine which table and mapping belong to each of the given relations. Explain if relation U and V are functions.

U: {(3, 4), (5, 2), (−2, 1), (4, 6), (3, −5)}   V: {(−5, 3), (−2, −1), (−1, 2), (0, 0), (2, 0)}

x	y
−2	1
3	4
3	−5
4	6
5	2

x	y
−5	3
−2	−1
−1	2
0	0
2	0

Relation U is _____ because the element 3 _____ in the domain.

Relation V is _____ because the domain values _____.

☑ **Checkpoint**

Create a table and a mapping for the relation. Explain if the relation is a function.
P: {(−12, 8), (−8, −8), (4, −8), (8, 8)}

EXPLORE 7A

▶ The Vertical Line Test

- The _____ compares a vertical line to every *x*-value on a graph.

- If you move a vertical line across a graph and it touches _____ _____, then the equation is a function.

- If you move a vertical line across a graph and it touches _____ _____ for any *x*-value, then the equation is not a function.

- _____ lines are not functions because every *x*-value has the same coordinate.

- Functions can be identified using a _____, _____, _____, and _____.

Example 6

Determine if the graphs below are functions.

A) 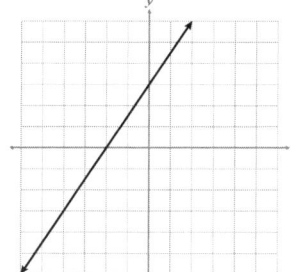   B) 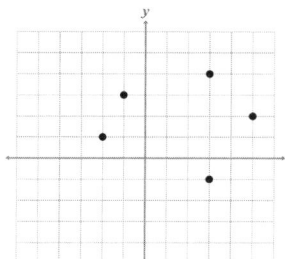   C)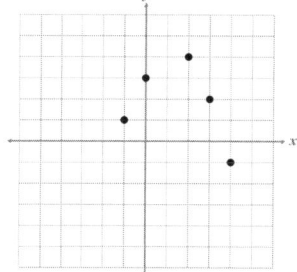

☑ Checkpoint

Determine whether the following relations are functions. Explain.

A)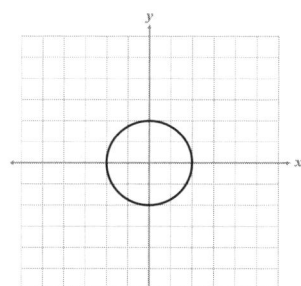

B)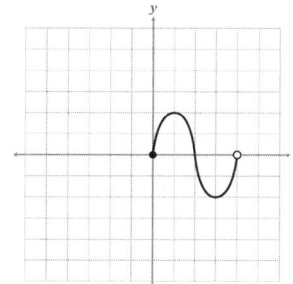

7A PRACTICE 1

Practice 1

Complete the problems on a separate sheet of paper.

1) On a coordinate plane, create rectangle $CDEF$: $\{C(-1, 5), E(2, -4), F(2, 5)\}$. Then find point D.

2) Name one vertical line segment and one horizontal line segment from problem 1.

3) List the domain and range of the relation.
$R: \{(-4, 8), (-2, 6), (0, 4), (2, 2), (4, 0)\}$

4) Reorganize $R: \{(-4, 8), (-2, 6), (0, 4), (2, 2), (4, 0)\}$ as a table.

5) List the domain and range of the relation.
$Q: \{(-1, -3), (0, -2), (1, -1), (1, 3), (2, 0)\}$

6) Reorganize $Q: \{(-1, -3), (0, -2), (1, -1), (1, 3), (2, 0)\}$ as a table.

Given the graph of the relation, complete the following:

A) Determine if the relation is a function. Explain.
B) List the set of points that make up the relation.

7)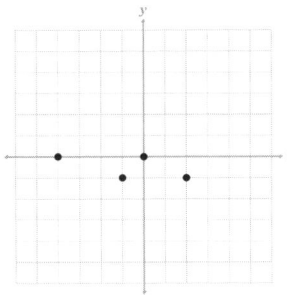

8)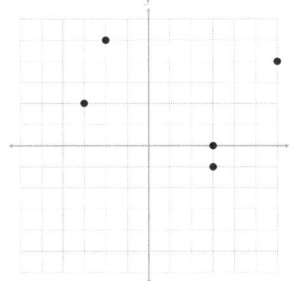

9) Name which problem (7 or 8) represents a function. Write the domain and range.

10) Given an ordered pair (x, y), which coordinate is a domain value? What are two other names for the domain value?

11) How can you tell that a relation is a function from a set of ordered pairs or a table?

Given a table, graph, or mapping, determine if the relation is a function. If a function is present, write the domain and range.

12)

x	y
-2	2
-1	1
$-\frac{1}{2}$	$\frac{1}{2}$
0	0
$\frac{1}{2}$	$\frac{1}{2}$
1	1
2	2

13)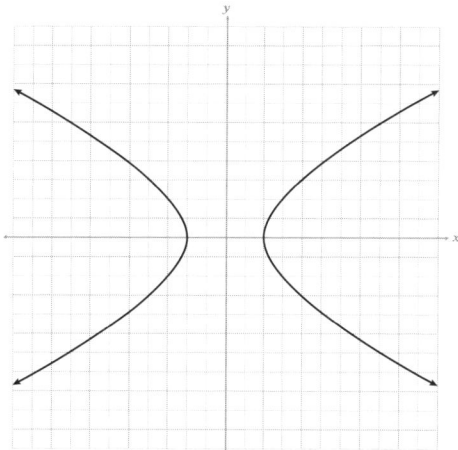

14)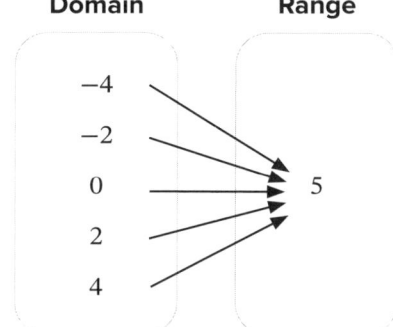

7A   MASTERY CHECK

✓ Mastery Check

✎ Show What You Know

A) What is important about the domain of a function?

Use the integers −9 through 9. You may use each number more than once in parts B and C.

B) Create a relation as a *table* that *is* a function.

x	y

C) Create a relation as a *mapping* that is *not* a function.

Domain   Range

D) If you combine your relation in part B and part C, will this be a function? Explain.

🔊 Say What You Know

In your own words, talk about what you have learned using the objectives for this part of the lesson and your work on this page.

Practice 2

Complete the problems on a separate sheet of paper.

1) On a coordinate plane, create rectangle $PQRS$: $\{Q(4, -3), R(4, 3), S(-1, -3)\}$. Then find point P.

2) Name both vertical line segments and horizontal line segments from problem 1.

3) List the domain and range of the relation.
$R: \{(-1, 0), (0, 4), (2, -1), (3, 0), (5, -2)\}$

4) Create a mapping of the relation.
$R: \{(-1, 0), (0, 4), (2, -1), (3, 0), (5, -2)\}$

5) List the domain and range of the relation.
$P: \{(-5, 4), (-2, -3), (-1, -3), (-1, 6), (0, 5)\}$

6) Create a graph of the relation.
$P: \{(-5, 4), (-2, -3), (-1, -3), (-1, 6), (0, 5)\}$

Given the graph of the relation, complete the following:

A) Determine if the relation is a function. Explain.
B) List the set of points that make up the relation.

7)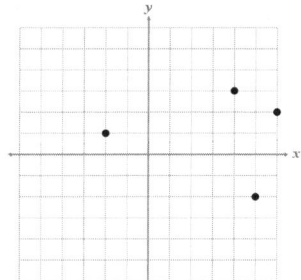

8)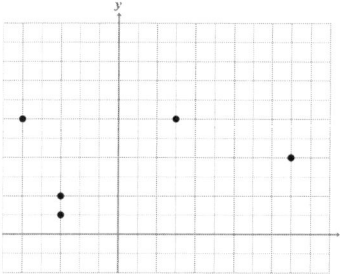

9) Identify which problem (7 or 8) represents a function, then write the domain and range.

10) Given an ordered pair (x, y), which coordinate is a range value?

11) Describe how you can tell if a relation is a function from a mapping.

7A PRACTICE 2

Given a table, graph, or mapping, determine if the relation is a function. If a function is present, write the domain and range.

12)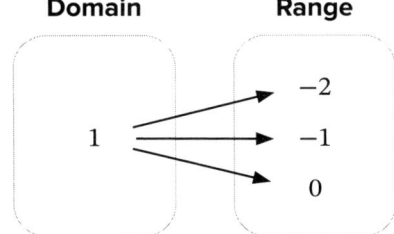

13)

x	y
−4	−2
−2	0
0	2
$\frac{1}{2}$	$\frac{5}{2}$
2	4

14)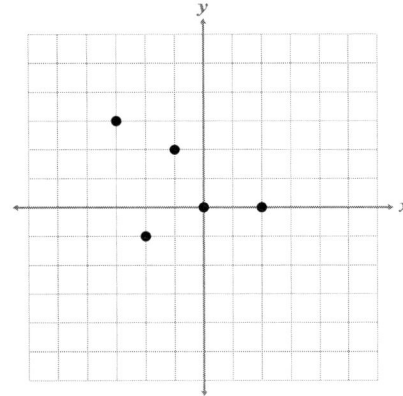

Part B: Understanding Functions

Objectives

In this part of the lesson, you will learn about understanding functions.

By the end of this lesson, you will be able to do the following:

- Write equations with two variables in function notation.
- Identify variables as dependent or independent.
- Evaluate a function for the dependent variable knowing a set of values for the independent variable.
- Determine whether a specific point is a solution for a function presented as a graph, table, or equation.

Why?

Determining the independent and dependent variables and how they relate to a function in function form is a critical part of creating the Plan when problem solving.

Warm Up

Substitute 4 for x. Do not simplify or solve.

1) $f(x) = -2x + 3$

2) $g(x) = 13x - 7$

Function Notation

- _____, $f(x)$, represents a function, f, where x is an input value in the function's domain and $f(x)$ is an output value in the function's range.

- Ordered pairs in function notation are written as _____.

- Function notation simply replaces _____ with $f(x)$.

- Function notation is useful when determining the _____ between dependent and independent variables.

Example 1

Write the equations in function notation with f in respect to x.

A) $y = 3x + 7$

$f(x) = 3x + 7$

B) $y = \frac{1}{2}x - 5$

7B EXPLORE

Example 2

Write the equations as y in terms of x.

A) $g(x) = 7x - 1$
   $y = 7x - 1$

B) $h(x) = -8x + 2$

☑ Checkpoint

Write the given equation in function notation using $f(x)$.

$y = -\frac{2}{3}x$

▶ Independent and Dependent Variables

- In a function, x is the _____ because x does not depend on the other variable for its value.

- The value $f(x)$ is dependent upon the value of x and, therefore, is the

  _____.

x	y

- The question you should ask yourself when you see a word problem is:

- Defining the variables as an ordered pair gives you a guide to how _____ relate to one another.

- The (_____ variable phrase) *depends* on the (_____ variable phrase).
       y x

Example 3

Define your variables as an ordered pair in words. Write a sentence that connects the independent and dependent variables.

A) Number of people attending and the total cost

   Ordered pair in words: (number of people, total cost)

   Sentence: The <u>total cost</u> *depends* on the <u>number of people attending</u>.

B) Distance driven and time in the car

   Ordered pair in words:

   Sentence:

☑ **Checkpoint**

Determine the dependent variable and the independent variable in the function below. Write the relationship as an ordered pair in words and then as a sentence.

Hourly earnings and time worked

▷ Using a Function Rule

- Evaluating an equation in function notation uses _____.

- An equation in function notation is also called a _____.

- Later in this unit, you will learn to graph a function on the coordinate plane. A table is one way to help you graph ordered pairs for $(x, f(x))$.

Example 4

Given the function rule, complete the table.

$f(x) = -3x$

Replace x with the values from the table, one at a time.
Evaluate the right side of the equation.

$f(-1) = -3(-1)$ $f(1) = -3(1)$
$f(-1) = 3$ $f(1) = -3$

$f(0) = -3(0)$ $f(2) = -3(2)$
$f(0) = 0$ $f(2) = -6$

Place the resulting value into the table in the $f(x)$ column.

x	$f(x)$
-1	
0	
1	
2	

7B EXPLORE

- Once you can determine the independent and dependent variables and use an equation in _____, you are ready to start _____ this to word problems.

- Using _____ values to complete the function rule or _____ to determine if a given ordered pair is part of the function are some of the ways you will _____ with function rules.

Example 5

A trampoline park charges a $15 admission fee for each person to enter.
The park uses the equation $c(p) = 15p - 10$ to determine the cost for groups, including a $10 discount.

Determine the ordered pair in words for the independent and dependent variables. Then determine the cost for $p = \{10, 12, 25, 32\}$.

The cost depends on the number of people attending.

Ordered pair in words:

One group attending says the cost for 20 people will be $200, or $c(20) = 200$. Is this correct for the given equation?

$c(20) = 15(20) - 10$
$c(20) =$

The group is incorrect. The cost would be _____, not $200.

☑ Checkpoint

Determine how much money Jim will make when he works for 5 hours, 8.5 hours, and 20 hours.

Jim works at a pizza shop and makes $9 per hour. He uses the equation below to determine his paycheck before any deductions.

$p(h) = 9h$

Practice 1

Complete the problems on a separate sheet of paper.

1) What can you replace "y" with when writing an equation in function notation? Consider the ordered pair (x, y).

Write the following equations in function notation where $(x, f(x))$.

2) $y = \frac{3}{4}x - 2$

3) $y = -x$

Write the following equations in function notation where $(x, g(x))$.

4) $y = 10x$

5) $y = \frac{5}{8}x + \frac{17}{8}$

Write the following equations as y in terms of x.

6) $g(x) = -6x + 11$

7) $h(x) = \frac{4}{3}x - 2$

For problems 8–9, complete the following:
  A) Determine the independent and dependent variables for the given context.
  B) Write as an ordered pair in words.
  C) Write a sentence comparing the dependent and independent variables.

8) The number of people attending a concert (p) and the amount of money collected at a concert (m).

9) The rate (r) water is running and the amount of water (w) used.

Given the domain: $\{-1, 0, 1, 2\}$, find the range of the function. Show your work. Record the ordered pairs in a table.

10) $g(x) = -\frac{1}{2}x + 1$

11) $h(x) = 5 - x$

---

12) Make a table when x is $\{-8, 0, 1, 8\}$ for the functions $f(x) = \frac{5}{8}x + \frac{17}{8}$.

13) Find $f(2)$, $f(-1)$, $f(5)$ when $f(x) = 3x - 2$. Is $f(3) = 9$ a solution to this function? Explain.

14) Find $g(2)$, $g(-1)$, $g(5)$ when $g(x) = x^2$

15) Camryn is running a race. He runs an average of 5 miles per hour.

  A) Define your variables as an ordered pair in words.

  B) Use the equation $d(t) = 5t$ to evaluate when $d(0.5)$, $d(1.5)$, $d(2)$.

7B MASTERY CHECK

Mastery Check

Show What You Know

The Ross family uses the equation $c(v) = 75v + 200$ to determine how much to save for their meals on each vacation.

A) The two variables the Ross family are comparing are the number of days on vacation and the cost of meals. Complete the sentence:

_____ depends on _____.

B) Complete the table to determine the cost for $v = \{2, 5, 7, 10\}$.

v	$c(v)$

C) Suppose on their last trip, the Ross family spent $500 on restaurant meals. How many days were they on their trip? Show your work.

D) The Ross family budgeted $3,000 for a 7-day trip to include all meals, gas, and excursions. If $725 was budgeted for food, gas costs the family $40 for each day of the trip and each excursion is $400, how many excursions will the family be able to schedule? Show your work.

Say What You Know

In your own words, talk about what you have learned using the objectives for this part of the lesson and your work on this page.

Practice 2

Complete the problems on a separate sheet of paper.

1) When $f(x)$ or $g(x)$ are written in an equation, what single variable can you replace them with? Is this the independent variable or the dependent variable?

Write the following equations in function notation where $(x, f(x))$.

2) $y = x - 3$

3) $y = \frac{2}{5}x + \frac{1}{2}$

Write the following equations in function notation where $(x, g(x))$.

4) $y = 1.25x + 26.50$

5) $y = -8x$

Write the following equations as y in terms of x.

6) $h(x) = x$

7) $f(x) = 7.25x$

For problems 8–9, complete the following:

A) Determine the independent and dependent variables for the given context.
B) Write as an ordered pair in words.
C) Write a sentence comparing the dependent and independent variables.

8) The hours you work (h) and the amount you get paid (p).

9) The velocity of a ball (v) and the angle (a) of an incline the ball is rolled down.

Given the domain: $\left\{-4, \frac{1}{2}, 8\right\}$, find the range of the function. Show your work. Then create a table.

10) $f(x) = 10x$

11) $f(x) = \frac{3}{4}x - 2$

---

12) Make a table when x is $\{-3, -2, -1, 0, 1\}$ for the functions $h(x) = \frac{x}{3}$.

13) Find $f(2), f(-1), f(6)$ when $f(x) = \frac{4}{3}x$

Which of the following ordered pairs are a solution to the function? Explain.
$(3, 7)$ or $(-3, -4)$

14) Find $g(0), g(4), g(6)$ when $g(x) = \sqrt{x}$

15) Tristan was participating in a charity event by jumping rope. For this event, his grandpa promised to donate $0.50 (d)$ to the charity every time Tristan jumped the rope (r).

A) Define your variables as an ordered pair.

B) Use the equation $d(r) = 0.50r$ to determine how much money Tristan will raise if he jumps: 50 times, 100 times, and 275 times.

C) Tristan says that if he jumps rope 1,000 times he will earn $5,000. Explain if Tristan is correct.

TARGETED REVIEW 7

Targeted Review

In the Targeted Review, you will practice topics you have mastered in earlier lessons. Reviewing these concepts will help you be successful as you work through this unit.

Complete the problems on a separate sheet of paper.

1) Twice the difference of a number minus seven is the same as three-fourths the number, plus seven halves. Write an equation and solve for the unknown value, n.

2) Solve and graph the solution(s) on a number line.
$-\frac{1}{2}x - 4 \geq 3$ OR $\frac{1}{3}x + 5 > 2$

3) Convert 72 cm² to in². Round to the nearest unit. (1 in = 2.54 cm)

4) Solve. $\frac{3}{8} = \frac{x}{2x + 3}$

Use the graph to answer problems 5 and 6.

5) For points A–F, write the ordered pair for each of the given points. Then name the quadrant to which it belongs.

6) For the ordered pairs {(0, 6), (−2, 0), (−4, 4), (2, −4), (0, 0), (−4, −4)}, name the point on the graph and the quadrant to which it belongs.

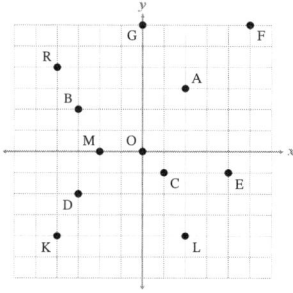

Complete the sentence with one of the following: always, sometimes, never.

7) Compound inequalities _____ use the word AND or OR.

8) The absolute value of x will _____ be less than zero.

---

9) Calculate the five-number summary for the data set. {23, 25, 21, 20, 24, 26, 30, 29, 27}

10) Determine if there are any outliers for the data set in problem 9.

Write the equation in terms of y.

11) Write the equation in terms of y.
$5x - \frac{2}{3}y = -4$

Multiple Choice

_____ 12) The mean of a data set is 75. The element 8 is added to the data set. How will this affect the mean?

   A) The mean will decrease.
   B) The mean will increase.
   C) The mean will stay the same.
   D) Not enough information.

_____ 13) The ratio of students that ride the bus to students who drive to school is 8 to 3. If there are a total of 1,100 students at the school, how many students drive?

   A) 3
   B) 300
   C) 413
   D) 800

Lesson 8
Using Graphs

Outline

Part A
Intercepts and Slope from a Graph

- The Intercepts
- Slope
- Graphing from the Slope and a Point

Part B
Translating the Linear Parent Function

- Representing the Linear Parent Function
- Translations of the Linear Parent Function

Targeted Review

Vocabulary

- y-intercept
- x-intercept
- slope
- parent function

8A EXPLORE

Part A: Intercepts and Slope from a Graph

Objectives

In this part of the lesson, you will learn about intercepts and slope.

By the end of this lesson, you will be able to do the following:

- Identify intercepts of a line from a graph, table, or equation.
- Given a graph of a line, determine the slope using slope triangles or rise over run.
- Plot the graph of a line given a point and the slope.

Why?

Slope and rate of change are key concepts in Algebra 1. Reading graphs and picking out key details like the x- and y-intercepts and the slope will help you solve problems in later lessons.

☁ Warm Up

Determine if the given ordered pair is a solution for the function.

1) $y = -2x + 3; (1, 1)$

2) $3x - 4y = -12; (0, -4)$

3) Explain how you determined your answers for problems 1 and 2.

▶ The Intercepts

- Each _____ on a coordinate plane represents an input value from the x-axis and a matching output value from the y-axis.

  - The _____ of an ordered pair determines proper placement on the coordinate plane.

EXPLORE 8A

- The intercepts of a graph are points that _____ the axes of the coordinate plane.

    - The _____-intercept is where the value of x is equal to _____.

        - The letter _____ is used to represent the y-intercept.

        - The y-intercept can be written as _____ or _____.

    - The _____-intercept is where the value of y is equal to _____.

        - The letter _____ is used to indicate the x-intercept.

        - The x-intercept can be written as _____ or _____.

Example 1

Determine the x-intercept and y-intercept for the function represented on the graph below. Mark the points and label them on the graph.

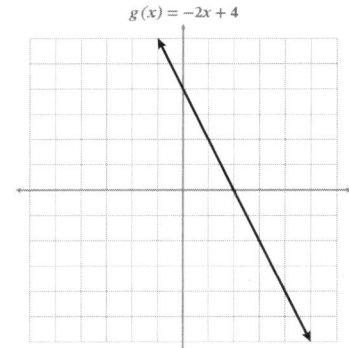

$g(0) = 4$, so $b = 4$

The y-intercept of this line is 4, which represents the ordered pair $(0, 4)$.

$g(2) = 0$, so $a = 2$

The x-intercept of this line is 2, which represents the ordered pair $(2, 0)$.

Example 2

Determine the x-intercept and y-intercept for the function represented in the table.

Plan Find the zeros in the table to get started.

Implement

The x-intercept of this line is _____.

$f(\boxed{}) = \boxed{}$, so $a = -2$

The y-intercept of this line is _____.

$f(\boxed{}) = \boxed{}$, so $b = -1$

x	$f(x)$
-2	0
-1	$-\frac{1}{2}$
0	-1
2	$-\frac{3}{2}$
2	-2

8A EXPLORE

Example 3

Determine the *x*-intercept and *y*-intercept for the function $h(x) = 2x - 3$.

Plan Find the *x*-intercept by substituting a for the value of x and 0 for the value of $h(x)$.

Then solve for a.

Find the *y*-intercept by substituting 0 for the value of x.

Then solve for b.

Implement

x-intercept:
$$h(a) = 0, \text{ or } (a, 0)$$

y-intercept:
$$h(0) = b, \text{ or } (0, b)$$
$$h(0) = 2(0) - 3$$

$$h\left(\frac{3}{2}\right) = 0$$

The *x*-intercept of this line is _____. The *y*-intercept of this line is _____.

☑ Checkpoint

Determine the *x*-intercept and *y*-intercept for the function represented below.

$f(x) = 3x + 4$

EXPLORE 8A

▶ Slope

- Slope of a line is the _____ ratio between any two points on a line.

    - Slope is represented by the letter _____.

- A line represents a function that changes at a _____ rate.

- In this course, when asked to find the slope, the answer should be a _____ value written as a _____ ratio.

- Two ways to think of slope visually are:

- To determine the sign of the slope before finding its numerical value, always read the graph from _____.

8A EXPLORE

Four Types of Slopes

Zero Slope

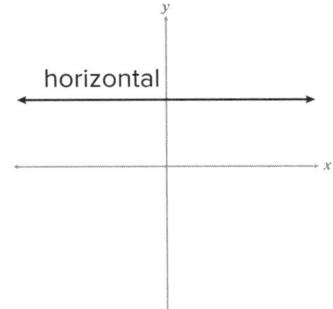

- As the *x*-values _____,

  the *y*-values _____.

- If the line is horizontal (—), then the range value does not change,

  so the slope is _____.

Negative Slope

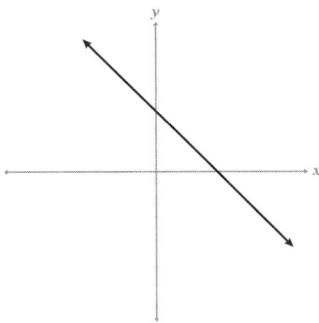

- As the *x*-values _____,

  the *y*-values _____.

- If the line slopes down and to the right (\ ), then the range values are

  decreasing, so the slope is _____.

Positive Slope

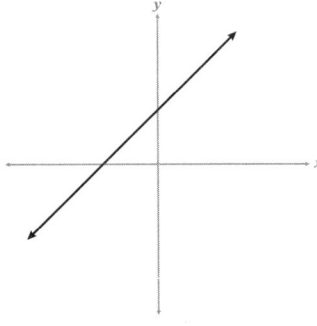

- As the *x*-values _____,

  the *y*-values _____.

- If the line slopes up and to the right (/), then the range values are

  increasing, so the slope is _____.

Undefined Slope

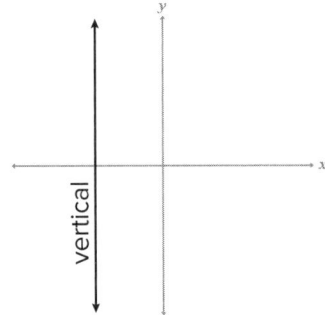

- As the *x*-values _____,

  the *y*-values _____.

- If the line is vertical (|), then the slope is _____.

  (This type of line is not a function.)

EXPLORE 8A

Example 4

Determine the slope from the graph. Explain the slope.

Plan Mark the two points you chose on the graph.

Start at the point on the left side and work your way across the graph.

Then record the number of spaces you moved up (positive direction) or down (negative direction) and the number of spaces you moved from left to right.

Write the slope as a ratio set equal to m.

Implement

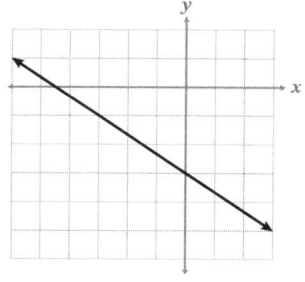

Explain

As the x-values increase, the y-values _____ .

Example 5

Determine the slope from the graph. Explain the slope.

Implement

This line has a _____ slope.

$m =$

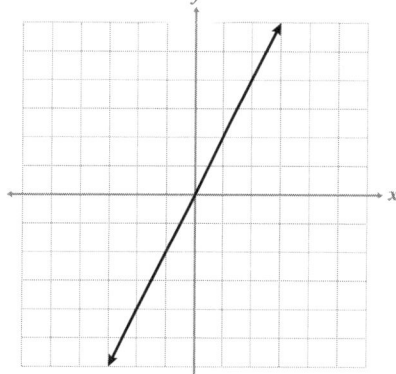

Explain
When you move from left to right, as the x-values increase, the

y-values _____ .

> The slope between *any two points on the same line* is equal.

8A EXPLORE

☑ Checkpoint

**Determine the slope from the graph.
Mark your points.**

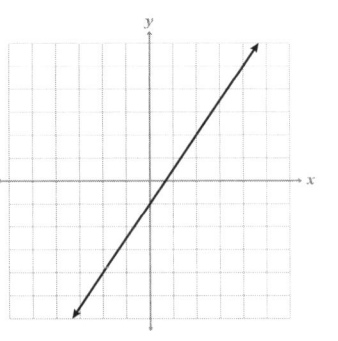

▶ Graphing from the Slope and a Point

- Many lines can be drawn through a single point, but _____ line can be drawn through any two points on the coordinate plane.

- If you have the _____ and *one point* on a graph, you have enough information to graph the line.

Example 6

Graph the line given the slope and a point when $m = -\frac{1}{2}$ and $(-2, 3)$ create the graph of line h.

Plan Mark the given point on the coordinate plane.
Then use the given slope to find one point in either direction.

Implement

Mark the point $(-2, 3)$ on the coordinate plane.

Use the slope $-\frac{1}{2}$ to move down 1 unit and right 2 units.

Then, move up 1 unit and left 2 units.

Connect all points with a line.

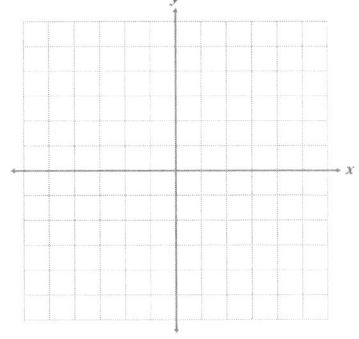

☑ Checkpoint

Graph a line given a point and the slope.

$(-1, -2), m = \frac{2}{3}$

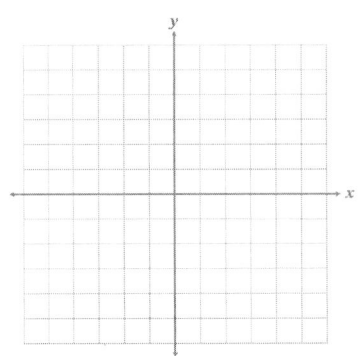

154 Lesson 8: Using Graphs > Part A: Intercepts and Slope from a Graph > Explore

Practice 1

Complete the problems on a separate sheet of paper.

Find the *x*-intercept and *y*-intercept for each function, and write them as ordered pairs.

1)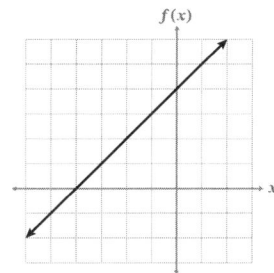

2)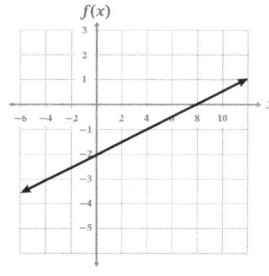

3)

x	$f(x)$
−1	−4
0	−2
1	0
2	2

4)

x	$g(x)$
−3	0
0	1
3	2

5) $h(x) = x + 2$

6) $f(x) = 3x - 1$

Determine the slope of $f(x)$.

7) Is the slope positive, negative, zero, or undefined? Explain.

8) Mark the slope of line $f(x)$ on the graph.

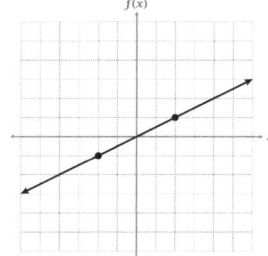

Determine the slope of line n.

9) Is the slope positive, negative, zero, or undefined? Explain.

10) Explain what undefined slope means.

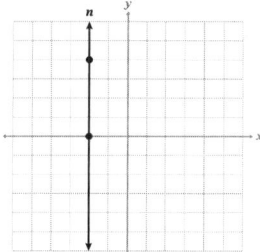

Given a point and the slope, graph the line on a coordinate plane. Then, mark the point to the left and right of the given point.

11) $(1, 4)$, $m = -\frac{1}{3}$

12) $(0, -3)$, $m = \frac{2}{5}$

Graph the linear equation on a coordinate plane.

13) $f(x)$: $(3, 1)$ and $m = \frac{3}{2}$

14) $h(x)$: $(0, 4)$ and $m = -1$

15) Suppose you needed to find the next point but you do not have a graph. Determine the next point to the right of the given point.
$f(x)$: $(3, 1)$ and $m = \frac{3}{2}$

8A MASTERY CHECK

Mastery Check

Show What You Know

Marco and Nick were given the graph and asked to find the slope. Both marked the same two points shown on the graph below.

A) Marco says that the slope is $\frac{-4}{3}$. Nick says the slope is $\frac{4}{-3}$. Is it possible that they are both correct? Explain.

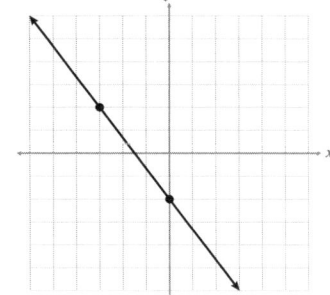

B) Marco and Nick are given the function $h(x) = \frac{2}{5}x - 2$ and need to determine the intercepts. Who has the correct intercepts for the function? Show your work.

$h(x) = \frac{2}{5}x - 2$

Marco	Nick
(5, 0)	(0, 5)
(0, −2)	(−2, 0)

C) How is it possible that both equations have the same y-intercept but are different lines?

Say What You Know

In your own words, talk about what you have learned using the objectives for this part of the lesson and your work on this page.

Practice 2

Complete the practice problems on a separate sheet of paper.

Find the x-intercept and y-intercept for each function below and write them as ordered pairs.

1)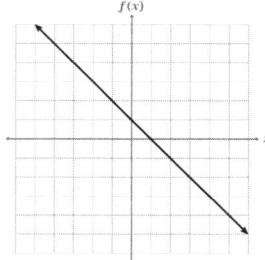

2) $g(x) = x - 7$

3)
x	f(x)
−3	−2
0	0
3	2

4)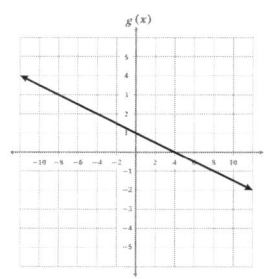

5)
x	h(x)
−1	2
$-\frac{1}{2}$	0
0	−2
$\frac{1}{2}$	−4

6) $f(x) = -\frac{1}{2}x + 4$

7) Is the slope of $g(x)$ positive, negative, zero, or undefined? Explain.

8) Find the slope of the graph $g(x)$.

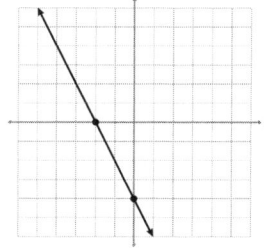

9) Determine the slope of $f(x)$.

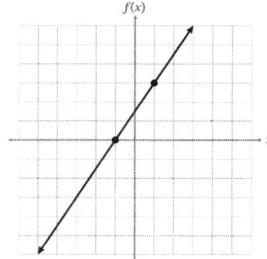

Graph a line given a point and the slope.

10) $h(x)$: $(-2, -3)$ and $m = 1$

11) $f(x)$: $(2, 0)$ and $m = -\frac{1}{3}$

12) $g(x)$: $(2, 0)$ and $m = -\frac{1}{3}$

13) $k(x)$: $(-4, 1)$ and $m = 4$

14) Describe a graph with a slope of zero. What is happening to the x- and y-values?

8B EXPLORE

Part B: Translating the Linear Parent Function

Objectives

In this part of the lesson, you will learn about the linear parent function and translations.

By the end of this lesson, you will be able to do the following:

- Express the linear parent function in all forms (table, graph, and equation).
- Demonstrate translations by a factor of b to a linear function.
- Explain how b translates a linear function up or down.

Why?

The linear parent function is the foundation for all other linear functions. Understanding this foundational graph will help you understand and compare all types of linear graphs throughout Algebra 1.

☁ Warm Up

Triangle A was moved to a new location on the coordinate plane (represented by triangle A').

1) How many spaces did the triangle move?

2) What direction was the triangle moved? Would this be considered positive or negative on the coordinate plane?

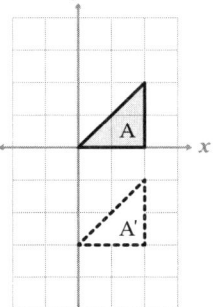

3) What is the y-intercept of triangle A'?

▶ Representing the Linear Parent Function

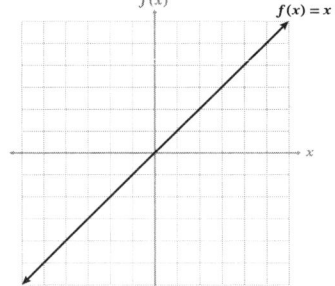

- The _____ is the simplest form of a function in the family of functions.

- The equation of the parent graph for the family of linear functions is _____ or _____.

- The graph of the linear parent function *always* travels through the _____.

- Both the y-intercept and the x-intercept of this line are zero, or _____.

- The slope of the linear parent function is _____ one.

EXPLORE 8B

Example 1

Complete the table for the parent function, $y = x$.

x	$h(x)$	(x, y)
-2	-2	
$-\frac{1}{3}$	$-\frac{1}{3}$	
0	0	
525	525	

> Since $y = x$, the value that is chosen for x will also be the y-value.

☑ Checkpoint

Given the parent function, $f(x) = x$, complete the table. What is the slope of the parent function?

x	$f(x)$
-20	
$-\frac{1}{2}$	
0	
11	

▶ Translations of the Linear Parent Function

- When you add or subtract any value to x, the _____ changes and translates the parent function.

- Adding a value to the function of any line will _____ the graph of that linear function up (if the number is _____) or down (if the number is _____) on the coordinate plane.

8B EXPLORE

Example 2

Complete the table and graph for $f(x)$ and $g(x)$. Then explain how the parent function is translated.

x	$f(x) = x$	$g(x) = x - 4$
−1		$g(-1) = -5$
0		−4
2		−2
4		0

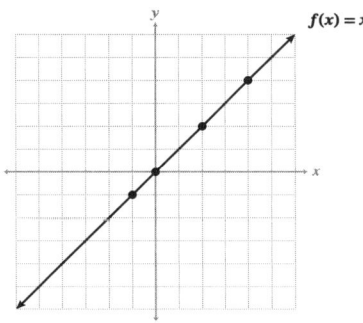

Explain

Adding _____ to the parent function (or _____ from the parent function), changes the y-value of $g(x)$ by _____. This means every point on the graph of $g(x)$ moves _____ units from the graph of $f(x)$.

Example 3

Complete the table and graph for $f(x)$ and $g(x)$. Then explain how the parent function is translated.

x	$f(x) = x$	$g(x) = x + 3$
−1	$f(-1) = -1$	
0	0	
2	2	
4	4	

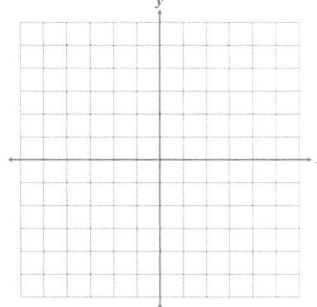

The graph is translated _____ 3 spaces from the _____.

☑ Checkpoint

How does changing the value of the y-intercept (b) affect the graph of $f(x) = x$? Describe when the y-intercept is positive and negative.

Practice 1

Complete the practice problems on a separate sheet of paper.

1) What is the simplest form of a function in a family of functions?
2) What is the x-intercept and y-intercept for the parent graph of a linear function?
3) Complete a table for the parent function when $x = \{-4, 0, 2, 5\}$.
4) Graph the parent function from your table in problem 3.
5) Would $f(7) = 8$ be a solution to the parent function? Explain.
6) How is it possible that the x-intercept and y-intercept are the same value for the parent function?
7) Complete a table for $f(x) = x$ and $g(x) = x + 1$ using the x-values $\{-4, 0, 2, 3\}$.
8) Explain how to transform the graph of $f(x)$ to create the graph of $g(x)$.
9) Graph $f(x)$ and $g(x)$ from problem 7 on the same coordinate plane.
10) Complete a table for $f(x) = x$ and $g(x) = x - \frac{1}{2}$ using the x-values $\{-2, -\frac{1}{2}, 0, \frac{1}{2}\}$.
11) Explain how to transform the graph of $f(x)$ to create the graph of $g(x)$.
12) Graph $f(x)$ and $g(x)$ from problem 10 on the same coordinate plane. Consider the scale of the graph.
13) Complete a table for $f(x) = 2x$ and $g(x) = 2x - 1$ using the x-values $\{-2, -1, 0, 1\}$.
14) Will the graph still translate by b when the coefficient of x changes?

8B MASTERY CHECK

Mastery Check

Show What You Know

A) Complete the table for $g(x) = \frac{3}{5}x$

x	$g(x)$
-5	
0	
5	

B) Graph the parent function $f(x) = x$ and $g(x) = \frac{3}{5}x$

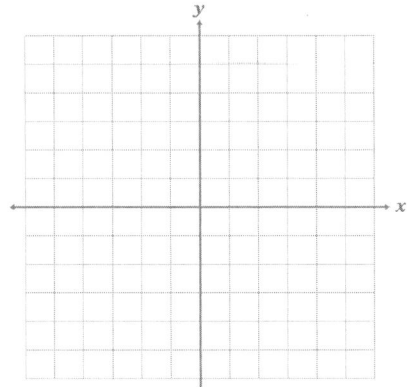

C) Will the slope or the y-intercept change between the parent function and $g(x)$? Explain.

D) Name the y-intercepts for the following functions.

	y-intercept
The function $h(x)$ is six spaces above $g(x)$, or $h(x) = g(x) + 6$	
The function $j(x)$ is two spaces below $f(x)$.	
$k(x) = g(x) - \frac{3}{4}$	

Say What You Know

In your own words, talk about what you have learned using the objectives for this part of the lesson and your work on this page.

Practice 2

Complete the practice problems on a separate sheet of paper.

1) What is the equation for the parent graph of linear functions in terms of f?

2) What is the slope for the parent graph of a linear function? Complete the table for the parent function.

x	$f(x) = x$
-2	
	0
$\frac{2}{5}$	
	56

3) Make a graph of the linear parent function. (Hint: Use some of the ordered pairs from the previous question.)

4) Harrison was given the function $h(x) = x - 20$. He believes that $h(-20) = -20$ is a solution to the translation of the parent graph. Explain why this is incorrect.

5) If the linear parent graph is shifted by the value of b (the y-intercept), will the graph go through the origin? Explain.

6) Complete a table for $f(x) = x$ and $g(x) = x - 5$ using the x-values $\{-4, 0, 2, 5\}$.

7) Explain how to transform the graph of $f(x)$ to create the graph of $g(x)$.

8) Graph $f(x)$ and $g(x)$ from problem 7 on the same coordinate plane.

9) Complete a table for $f(x) = x$ and $g(x) = x + 15$ using the x-values $\{-30, 0, 15, 40\}$.

10) Explain how to transform the graph of $f(x)$ to create the graph of $g(x)$.

11) Graph $f(x)$ and $g(x)$ from problem 10 on the same coordinate plane. Note the scale of the graph.

12) Complete a table for $f(x) = -\frac{1}{2}x$ and $g(x) = -\frac{1}{2}x + \frac{3}{2}$ using the x-values $\{-4, 0, 2, 5\}$.

13) Will the graph still translate by a factor of b when the coefficient of x changes?

14) What number can be added to the parent function and the graph still remain in the same location on the coordinate plane? Explain.

TARGETED REVIEW 8

Targeted Review

In the Targeted Review, you will practice topics you have mastered in earlier lessons. Reviewing these concepts will help you be successful as you work through this unit.

Complete the problems on a separate sheet of paper.

1) Explain what a function is in your own words. Use the words domain and range in your explanation.

Determine if the following ordered pairs are solutions to the graph of the function $g(x) = 3x$. Explain.

2) $(-2, -6)$

3) $(-2, 6)$

4) Determine the independent and dependent variables as an ordered pair. Then write a sentence that explains the relationship. Speed of a car and distance traveled

Label the domain and range in the table and mapping below. Determine if the given relation is a function. Explain.

5)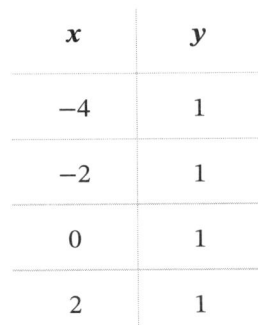

6)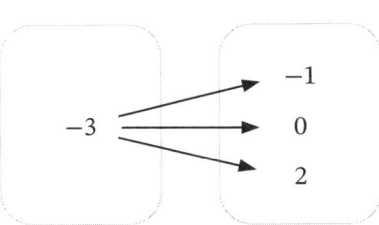

7) What is the name of the test that determines a function from a graph on the coordinate plane?

8) Solve. Justify each step with an algebraic property. $3(x - 4) = 5x + x - 7$

9) Graph the solutions for the inequality $-2p > 3$ on a number line.

10) When does the direction of the inequality symbol change?

11) Convert the following. Round to the nearest unit. 70 miles/hr = _____ km/hr (1 mile ≈ 1.6 km)

12) Solve. $\frac{y - 5}{-6} = \frac{-3}{2}$

Multiple Choice

13) Which equation correctly represents the function $f(x) = \frac{7}{2}x - 6$ as y in terms of x?

   A) $x = \frac{7}{2}y - 6$

   B) $2y = 7x - 12$

   C) $x = \frac{7}{2}f(x) - 6$

   D) $y = \frac{7}{2}x - 6$

14) Select *all* true answers for $\pi + 4$.

   ☐ real number
   ☐ rational number
   ☐ irrational number
   ☐ whole number

Lesson 9
Slope and Linear Equations

Outline

Part A
Slope and Graphed Scenarios

- Calculating Slope
- Rate of Change
- Describing and Sketching Graphs

Part B
Point-Slope Form and Slope-Intercept Form

- Point-Slope Form
- Point-Slope Form from Context
- Slope-Intercept Form
- Slope-Intercept Form from a Graph

Targeted Review

Vocabulary

- slope formula
- point-slope form
- slope-intercept form

9A EXPLORE

Part A: Slope and Graphed Scenarios

Objectives

In this part of the lesson, you will learn about slope and graphed scenarios.

By the end of this lesson you will be able to do the following:

- ✓ Use the slope formula, $m = \frac{\Delta y}{\Delta x}$, to calculate the slope of a line when given two points on the line.
- ✓ Describe a graph as a scenario using mathematical vocabulary and sketch a graph from a written scenario.

Why?

One of the most important things you will do in Algebra 1 is learn to use the slope formula. Linear equations that you will encounter throughout algebra are dependent on a solid understanding of slope.

Warm Up

1) How do you determine slope from a graph?

2) Why should you *not* use a graph to determine the slope of the line for the points $(-100, 234)$ and $(53, 3{,}412)$.

▶ Calculating Slope

- The slope of a line is _____ between any two points on the line.

- Finding the slope between any two points provides the slope for the _____.

- If the ordered pairs are provided in a table, you can find the slope by finding the

  _____.

- Remember that the slope is the _____ over the _____ (rise over run).

- Organizing points in a _____ is one way to help ensure that the order of the coordinates is correct.

EXPLORE 9A

Example 1

Find the slope from the given table.

x	y
−2	−3
4	5

$$m = \frac{\Delta y}{\Delta x} = \frac{8}{6} = \frac{4}{3}$$

- The slope formula is the _____ way to find the slope when given _____.

- Write the formula for slope:

- The symbol Δ is the capital Greek letter delta and means "_____."

- The subscripts represent different points: point one is (_____) and point two is (_____).

- It is important that the coordinates of each point align _____ in the slope formula.

Example 2

Find the slope of the line that includes the points (−2, −3) and (4, 5).
Label the provided information.

$$m = \frac{\Delta y}{\Delta x} = \frac{y_2 - y_1}{x_2 - x_1} = \frac{5 - (-3)}{4 - (-2)} =$$

- When using the slope formula, it is important to remember that the formula _____ the values from one another.

- If negative numbers are substituted into the formula, you may be able to _____ the values depending on which point is labeled (x_1, y_1) or (x_2, y_2). This is because subtracting a _____ is the same as adding its _____ value.

- To use the slope formula to find a missing coordinate, you find the _____ once the known values are substituted into the slope formula.

9A EXPLORE

Example 3

Find the value of r.
$m = -\frac{6}{5}$, $(-6, 8)$ and $(4, r)$

Plan Label the given points.
Substitute all known values into the slope formula.

Implement

Point 1 $(-6, 8)$

Point 2 $(4, r)$

$$\frac{m}{1} = \frac{y_2 - y_1}{x_2 - x_1}$$

$$\frac{-6}{5} = \frac{r - 8}{4 - (-6)}$$

☑ Checkpoint

**Find the slope of the line passing through $(-6, 13)$ and $(3, 7)$.
Show your work.**

▶ Rate of Change

- When a problem asks for the _____, it is usually referring to the slope.

- Using these strategies can be helpful when solving rate of change problems:

    - _____

    - _____

- When asked to find slope, the answer should be a numerical value written as

    a _____.

- When asked to find the rate of change, the answer should include the value of the slope

    with _____ related to the problem.

- The context will relate to the _____ and _____ variables.

- rate of change =

---

168 Lesson 9: Slope and Linear Equations > Part A: Slope and Graphed Scenarios > Explore   Algebra 1 Student Worktext

Example 4

Plan Write an ordered pair in words.
Find the rate of change.
Answer in a sentence.

Implement

The ordered pair in words:

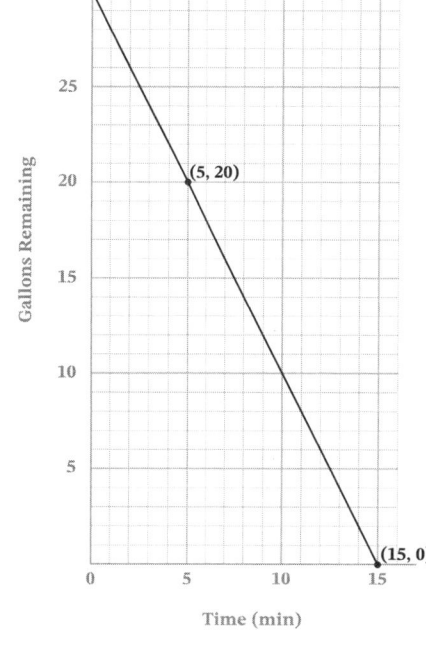

You can use the ordered pair in words to help make sense of the numbers.

- (0, 30) At _____ minutes, there are _____ gallons of water in the bathtub.

- (15, 0) At 15 _____, there are 0 _____ of water in the bathtub.

Rate of Change: $m = \frac{30 - 0}{0 - 15} = \frac{30}{-15} = -\frac{2}{1} = -2$

Explain

☑ Checkpoint

Define the variables as an ordered pair in words. Find the rate of change using the slope formula. Show your work.

Ethan drove 165 miles in 3 hours. Then it took him 2 hours to drive 110 miles. How fast is Ethan driving in miles per hour?

▷ Describing and Sketching Graphs

- Even on a non-linear graph, there can be parts of the graph that are _____.

9A EXPLORE

- Whether a graph is linear or nonlinear, each section of the graph can be described using words like these:

  - _____
  - _____
  - _____

Example 5

Describe each part of the graph.

The variables can be defined as

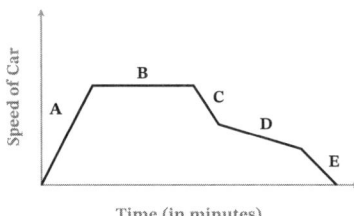

Trip to the Store

A	The speed of the car is _____ at a _____ rate.
B	The car reaches the speed limit and drives at a _____ for a few minutes.
C	
D	
E	

Example 6

Sketch a graph for the given scenario.

- You need to charge your cell phone.
- The battery is at 10% power when you start.
- You leave it plugged in overnight.
- The phone will charge at a constant rate until fully charged.

The variables are defined as

Cell Phone Charging

> When the graph is sketched, it should show what is happening in the given description.

Explain
The graph shows the battery starting slightly above the origin and increasing in percent charged until the phone is fully charged but still plugged in.

Example 7

Describe the graph.

Define the variables as an ordered pair:

Why did the ball start 4 feet above the ground (0, 4)?

Sam was holding the ball in his hand, _____ above the ground.

Why does the graph stop at (14, 0)?

_____

14 feet from where Sam was standing.

Explain

Sam was holding the ball _____ the ground and threw it into the air.

The ball reached a _____ of about _____ and landed on the ground _____.

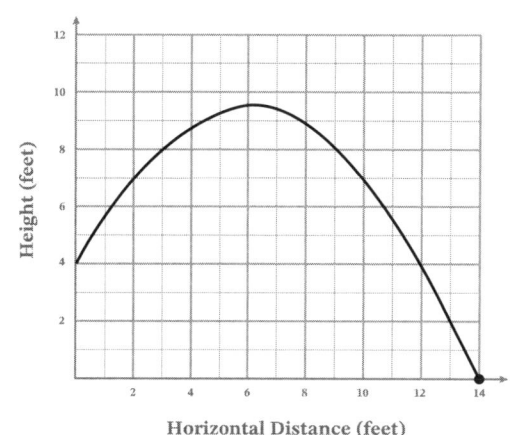

Sam Throws Ball

☑ **Checkpoint**

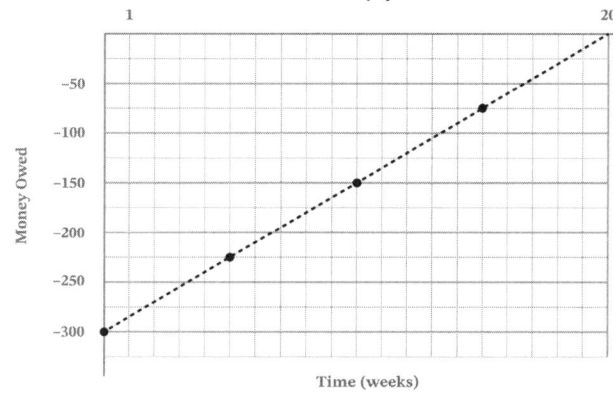

Loan Repayment

Joe borrowed $300 from his brother. They set up a weekly payment schedule. Joe's brother plotted the points on the graph to show the money still owed. Joe added the dashed line to determine if he could find a pattern.

A) Determine if the graphed scenario is linear or nonlinear.

B) Define the variables as an ordered pair.

C) Explain the meaning of the intercept(s).

D) If linear, determine the rate of change.

9A PRACTICE 1

Practice 1

Complete the problems on a separate sheet of paper.

1) Write the slope formula, then explain what each variable stands for.

Use the slope formula to determine the slope of the line.

2) $(8, 4)$ and $(-8, -4)$

3) $(4, -1)$ and $(-7, 3)$

4) $(-12, 14)$ and $(-10, 12)$

5) $(-15, 75)$ and $(-10, 50)$

Using the slope formula, find the missing value, r.

6) $m = \frac{4}{7}, (-5, 2), (9, r)$

7) $m = \frac{5}{2}, (3, 1), (r, 6)$

8) Josie's cell phone bill is $35 per month. Name the rate of change. Explain what it means in context and define your variables.

9) A three-year-old maple tree was 7 feet tall. When the tree was five years old, it had reached a height of 12 feet. How fast does the tree grow each year, assuming it grows at a constant rate?

Given the graph, describe the function by completing all of the following:

   A) Determine if the situation is linear or nonlinear.
   B) Define your variables.
   C) Explain the intercepts when possible.
   D) Explain the slope and rate of change when possible.

Answers may be estimates.

10)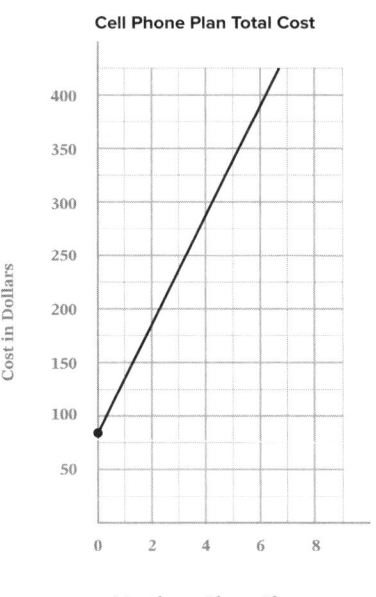
Cell Phone Plan Total Cost

11)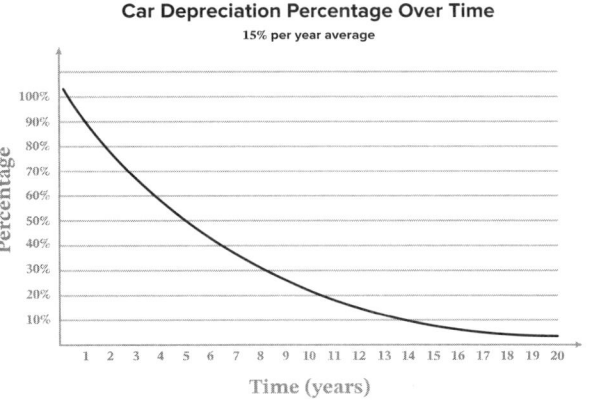
Car Depreciation Percentage Over Time
15% per year average

12) Oral Dosage Three Times Daily
6mg Total

Sketch a graph to provide a visual representation of the given scenario. Remember to label each axis, include a title, and label any points you think are important.

13) Sergio was standing at the edge of a cliff overlooking the ocean. He tossed a stone into the air and watched it fall into the water.

   Use the ordered pair (time, height above water) to sketch your graph.

14) Jude was mowing lawns for $20 per hour, and he had already saved $150.

   Use the ordered pair (hours, money) to help sketch your graph.

15) Malik started his run from the door of his house. He ran 1.5 miles away from his house and then turned around and ran back. Malik ran at a steady pace for his entire run, which took 30 minutes.

   Use the ordered pair (minutes, miles from house) to help sketch your graph.

9A MASTERY CHECK

📋 Mastery Check

✏️ Show What You Know

Rory was training for a 5-mile run. She did not run at the same pace the entire time and wanted to know what her rate was for different distances and time intervals.

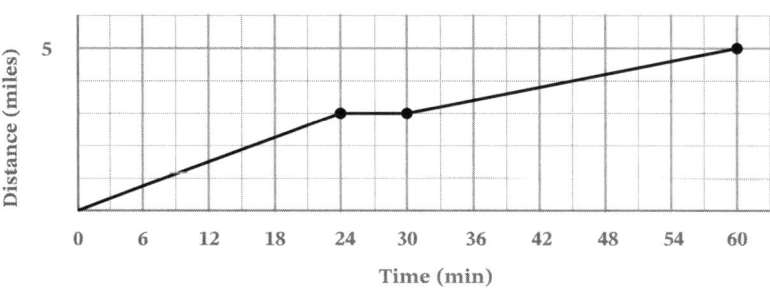

A) Explain the meaning of the ordered pairs (0, 0) and (60, 5).

Use the slope formula to help determine the rate of change from the graph.

B) What was Rory's pace for the first 24 minutes of the run?

C) When did Rory take a rest during the run? How long was the break? Use the slope formula to support your answer.

D) Write a scenario that describes the graph.

🔊 Say What You Know

In your own words, talk about what you have learned using the objectives for this part of the lesson and your work on this page.

Practice 2

Complete the problems on a separate sheet of paper.

Find the slope using the slope formula.

1) $(2, 0)$ and $(5, -1)$
2) $(-4, 1)$ and $(-3, 5)$
3) $h(x): (-2, -3)$ and $(-1, -2)$

Using the slope formula, find the missing value, r.

4) $m = -\frac{2}{3}, (-7, 2), (r, -4)$
5) $m = 8, (3, 2), (0, r)$

6) Suzy and Florence went to Milky Way Farm for milk sold by the ounce. Suzy's milk weighed 6 ounces and cost $2.88. Florence spent $6.72 for 14 ounces of milk. What is the cost per ounce of milk? Use the slope formula to solve.

7) Micah was running a marathon and wanted to find his average pace during the race. At the five-mile mark, Micah had been running for 55 minutes. After 220 minutes, he had run 20 miles. How many minutes does it take Micah to run one mile? Use the slope formula.

Given the graph, describe the function by completing all of the following:

A) Determine if the situation is linear or nonlinear.
B) Define your variables.
C) Explain the intercepts when possible.
D) Explain the slope and rate of change when possible.

8)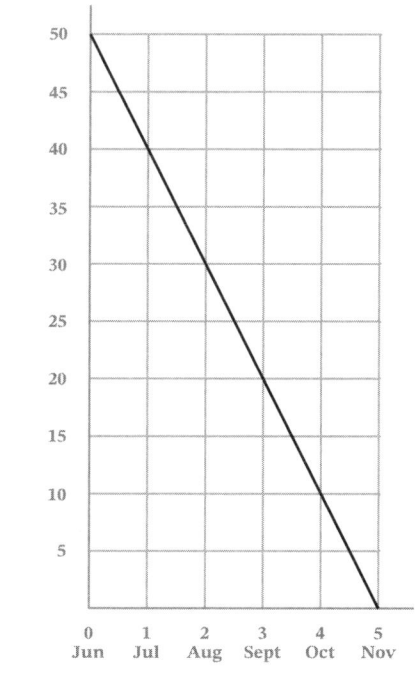

9)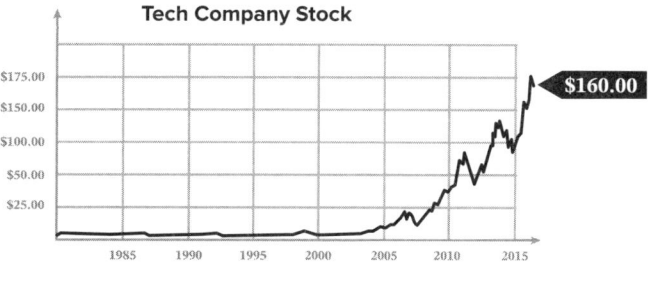

9A PRACTICE 2

Sketch a graph to provide a visual representation of the given scenario. Remember to label each axis, include a title, and label any points you think are important.

10) A sunflower was planted 1 inch below the ground and grew steadily to a maximum of 72 inches after 60 days. It stood tall for 10 days and then slowly drooped for 30 days until it fell to the ground.

   Use the ordered pair (days, height) to sketch the graph.

11) Etián started driving to the grocery store. He increased the speed of his car as he entered the highway. He drove at a constant speed until he exited the highway and had to slow down. He slowed down even more to enter the parking lot and park his car.

   Use the ordered pair (time, speed) to sketch a graph.

12) Georges was running a 10 kilometer race. He steadily increased his running speed for the first 2 kilometers until he found a constant pace for the next 5 kilometers. He started getting tired and slowed down for 1 kilometer and ran at about half speed for another 1.5 kilometers. For the last half a kilometer, he sprinted to the finish line.

   Use the ordered pair (running speed, distance ran) to sketch a graph.

13) Explain how creating a sketch of a scenario can help you make sense of a problem.

Part B: Point-Slope Form and Slope-Intercept Form

Objectives

In this part of the lesson, you will learn about point-slope form and slope-intercept form.

By the end of this lesson, you will be able to do the following:

- Write linear equations in point-slope form from a given graph or a point and the slope.
- Write linear equations in slope-intercept form from a graph or given the slope and the y-intercept.
- Graph equations on the coordinate plane in point-slope or slope-intercept form.

Why?

Being able to write linear equations in more than one form and represent them as an equation or graph is an integral part of Algebra 1. These skills will help you solve problems involving linear equations when given a variety of information about the line.

Warm Up

Substitute the values into the given equations: $A = 3$, $B = -\frac{3}{4}$, $C = \frac{1}{2}$

1) $Ax + By = C$

2) $y = A(x - B) + C$

Point-Slope Form

- Point-slope form is a way to write a _____ when provided the slope and the coordinates of any point on the line.

- Point-slope form is written as _____, where

    - (x_1, y_1) is a _____

    - m is the _____ of the line, and

    - _____ represent the independent and dependent variables.

9B EXPLORE

Example 1

Given the slope of line d is $-\frac{1}{3}$ and the point (6, 2) is on line d, find the point-slope form of the equation for line d.

Plan Identify m and (x_1, y_1)
Substitute what is known into the slope formula.
Write in point-slope form.

Implement

$\frac{-1}{3} = \frac{y_2 - 2}{x_2 - 6}$

$-1(x_2 - 6) = 3(y_2 - 2)$

$-\frac{1}{3}(x_2 - 6) = y_2 - 2$

$-\frac{1}{3}(x - 6) = y - 2$

Explain

◁ Substitute values into the slope formula.

◁ Find the cross product.

◁ Multiplication Property of Equality

◁ The remaining x and y represent any ordered pair that satisfies the equation of the line

◁ Point-slope form specifically has the y-values on the left side of the equation. The final solution must be in this form.

Example 2

Write the equation of a line in point-slope form when the slope is 2 and the line includes the point (3, −4).

Implement

$y - y_1 = m(x - x_1)$

$y - (\quad) = \quad (x - \quad)$

Explain

◁ Substitute values into the point-slope formula

◁ Simplify.

Example 3

Write the equation for the line that passes through the points (−3, 2) and (1, −6) in point-slope form.

Plan Find the slope using the slope formula.
Write the equation in point-slope form for the point (−3, 2).
Write the equation in point-slope form for the point (1, −6).

Implement

Point-slope form using the point (−3, 2) is

Point-slope form using the point (1, −6) is

Both equations represent the same linear equation but show different points on that same line.

Checkpoint

Write an equation in point-slope form where the slope is −3 and the line passes through the point (−6, 7).

Point-Slope Form from Context

- Point-slope form can be determined from a given context when the _____

  (the slope) and a _____ (point) is known.

Example 4

Write the equation in point-slope form for the scenario below.

Nathaniel had a cart where he sold flavored ice. He started his day off with some money in his cash box from the previous day's sales. He then earned $3 per cup of flavored ice that he sold. After selling 12 cups of flavored ice, Nathaniel had $76.

Plan (Independent variable, dependent variable)
Use the context to define the slope and a point on the line.
Use them to write the equation in point-slope form.

Implement

Any point on the line:

The rate of change is $3 per cup:

At one *specific* instance, Nathaniel sold 12 cups and had $76:

In point-slope form, the equation is

Checkpoint

The average cost for gas in 2019 was $2.47 per gallon. It cost $29.64 to fill a car with 12 gallons of gas.

Define your variables as an ordered pair. Write an equation in point-slope form.

9B EXPLORE

▶ Slope-Intercept Form

- Slope-intercept form is written as _____.

- Slope-intercept form can be used to identify the rate of change (or slope) and the _____ (y-intercept) of the function.

- In slope-intercept form, y is written in terms of x which means y is _____ on one side of the equation.

- Given the _____ and _____, you should be able to write the equation in slope-intercept form.

Example 5

Use the context to write the problem in slope-intercept form.

Liz hired a plumber to fix her washing machine. The plumber charged $50 per hour plus a one-time assessment fee of $80.

Plan Identify the variables

Figure out which value is the slope and which is the y-intercept

Implement

This equation could also be written in function notation: _____.

Maybe you defined your variables as something other than x and y like (h, c). In this case, you would write _____. All of these equations represent the _____ problem but using different notations.

☑ Checkpoint

Shelly purchased a car for $5,000. She paid $175 per month for car insurance and gas.

Write the equation in slope-intercept form to represent the cost of owning the car. Remember to define your variables as an ordered pair in words.

Slope-Intercept Form from a Graph

- You can write an equation in slope-intercept form from a graph by identifying the _____ or an _____ from a graph and find the _____ between points.

Example 6

Write an equation in slope-intercept form given line d.

Plan Mark the y-intercept of the graph. Mark another point on the line. Find the slope.

Implement

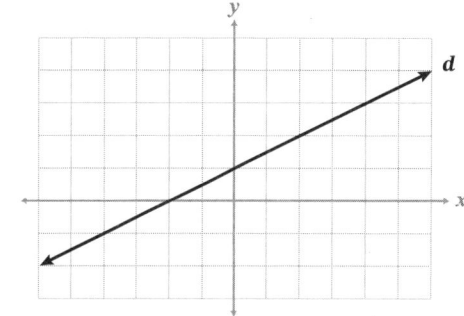

✓ Checkpoint

Write the equation in slope-intercept form for the function represented below.

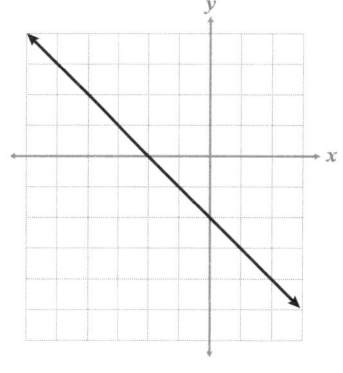

9B PRACTICE 1

Practice 1

Complete the problems on a separate sheet of paper.

1) Use the graph and marked point to write the equation in point-slope form.

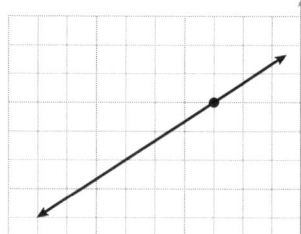

2) Given the slope of a line is $m = -1$ and a point on the line is $(0, 2)$.

   A) Write the equation in point-slope form.

   B) Use the equation to find the value of x when $y = 0$.

3) Given the points $(-7, -3)$ and $(-2, -5)$,

   A) Find the slope for this line.

   B) Write this line in point-slope form (using both points).

4) Given the points $(2, 3)$ and $(10, 9)$,

   A) Find the slope for this line.

   B) Write the equation of this line in point-slope form. Use the point $(2, 3)$.

Write an equation in point-slope form for the following scenarios. Define your independent and dependent variables as an ordered pair.

5) Mike sends an average of 15 texts per hour. After 8 hours, he had sent 120 texts.

6) Jessica works as a server for $12 per hour plus tips. She earned $150 after working an 8 hour shift.

Write in slope-intercept form.

7) $m = -\frac{3}{4}, b = 11$

8) $m = 5, b = -\frac{6}{7}$

9) $m = -1, (0, 9)$

10) $b = 2.5, m = 7.65$

11) $(0, 25), m = -\frac{5}{8}$

12) $m = 0, b = 3$

Write the equation in slope-intercept form from the graph.

13) Line a

14) Line b

15) Line c

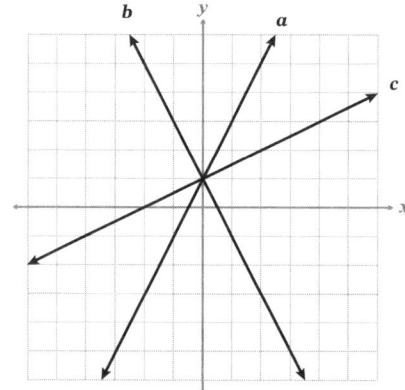

PRACTICE 1 9B

Name the slope and the point. Then graph the equation of the line on a coordinate plane.

16) $y - 2 = \frac{1}{3}(x - 4)$

17) $y + 3 = -4(x - 1)$

18) $y = \frac{3}{2}x - 4$

19) $y = -2x + 3$

20) Melissa has already saved $1,200 for a car. She plans on saving $200 each month until she can purchase a car. Identify the independent and dependent variables as an ordered pair. Name the value of m and b. Write an equation in slope-intercept form.

9B MASTERY CHECK

Mastery Check

Show What You Know

A tree was planted when it was 40 inches tall. It continued to grow 2.5 inches per year.

A) Define the variables as an ordered pair and define what the values in the given information represent.

B) Write an equation in slope-intercept form to represent the growth of the tree.

C) A seed was planted in the soil and grew at a rate of 1 inch every two days. After ten days, the plant was 4 inches tall.

Write an equation in point-slope form and graph your equation.

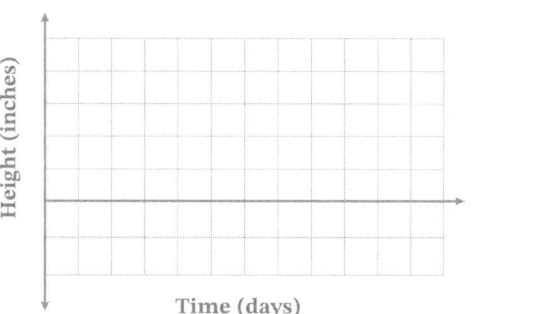

D) How far below the soil was the seed planted in part C? Explain. Write the equation in slope-intercept form for the graph in part C.

Say What You Know

In your own words, talk about what you have learned using the objectives for this part of the lesson and your work on this page.

Practice 2

Complete the problems on a separate sheet of paper.

1) Use the graph and marked point to write the equation in point-slope form.

2) Given the slope $-\frac{2}{3}$ and the point $(3, -2)$, write the equation in point-slope form.

3) Given the slope of a line is $m = \frac{3}{4}$ and a point on the line is $(-5, 1)$,

   A) Write the equation in point-slope form.

   B) Use the equation to find the value of y when $x = -1$

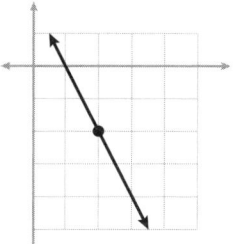

4) Given that the points $(0, 3)$ and $(-1, 5)$ are both on the same line,

   A) Find the slope for this line.

   B) Write this line in point-slope form (using either point).

   C) Using either equation, find the x-intercept by making $y = 0$.

Write an equation in point-slope form for the following scenarios. Define your independent and dependent variables as an ordered pair.

5) Rose sold bows for $5 each. After selling 25 bows, she had earned $125.

6) Jacob was a bricklayer and found that on average he laid 45 bricks per hour. After working 4 hours, he had 270 bricks laid for this project.

Use the graph of the line g to answer the following questions.

7) What is the y-intercept for this line?

8) If no exact values are provided, why can a graph only represent an estimate of the exact values for a function?

9) The equation for this graph is $y = -3x + \frac{3}{2}$. Explain the advantage of an equation of a function over the graph of a function.

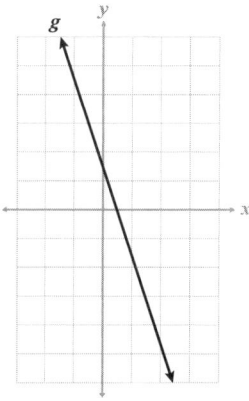

Write an equation in slope-intercept form that represents the graph.

10) $f(x)$

11) $g(x)$

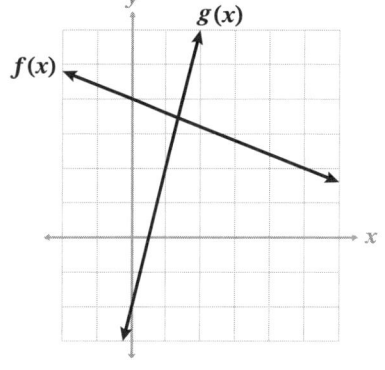

9B PRACTICE 2

Match the equation to the appropriate line provided.

12) $f(x) = -\frac{3}{2}x + 3$

13) $g(x) = -\frac{2}{3}x + 3$

14) $h(x) = 2\frac{2}{3}x + 3$

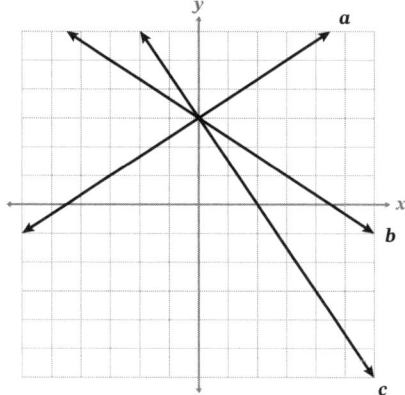

15) Lana drew line m on the graph. Allyson drew line p with twice the slope of line m and half the y-intercept of line m. Write the equation for line p in slope-intercept form. Graph line p on the coordinate plane.

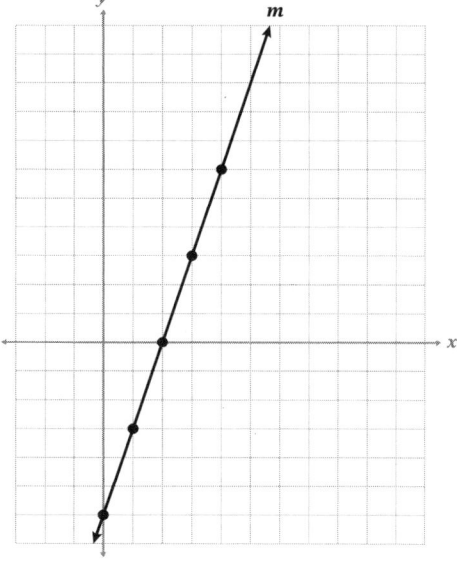

Name the slope and the point. Then graph the equation of the line on a coordinate plane.

16) $y - 4 = \frac{1}{2}(x + 3)$

17) $y = -3x + 2$

Bobbie noticed her electric bill had two parts, a fixed charge of $8.20 and a rate of $0.10 per kilowatt-hour (kWh).

18) Write an equation in slope-intercept form to represent the bill.

19) If Bobbie's electric bill was $106.20, how many kilowatt-hours did she use?

Targeted Review

In the Targeted Review, you will practice topics you have mastered in earlier lessons. Reviewing these concepts will help you be successful as you work through this unit.

Complete the problems on a separate sheet of paper.

1) Which graph represents the function $f(x) = 3x$? Explain. (Hint: Use $f(2) = 6$ to help explain your reasoning.)

2) Why would $f(-2)$ and $f(0)$ be poor choices for determining which graph represents $f(x)$?

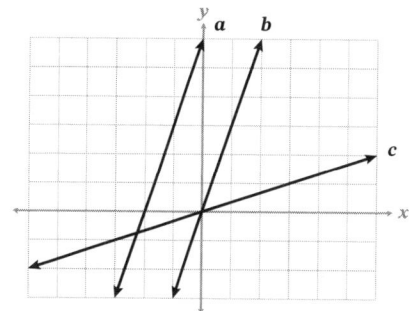

Use the graph to answer problems 3–6.

3) What are the intercept(s) of the graph?

4) Complete the table using the graph of the function.

x	y
−2	
	−2
0	
	2

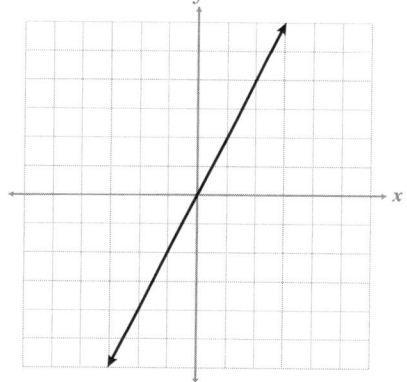

5) Determine the slope by marking the graph.

6) Name the domain and range using the table from problem 4.

---

7) Determine if the given relation is a function. Explain.

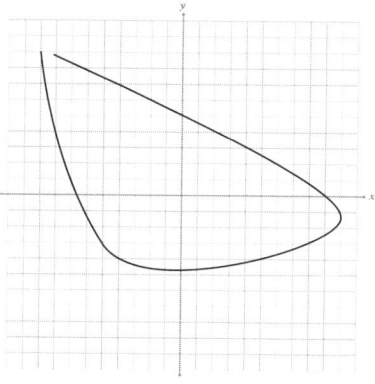

TARGETED REVIEW 9

8) As ordered pairs for the function f, do $(-2, 4)$ and $(4, -2)$ mean the same thing? Explain.

9) Solve for y. $6x - 5y = 2$

10) Solve for the interquartile range (IQR) of the following data set: $\{3, 7, 2, 0, 4, 3, 5, 3\}$

Multiple Choice

_____ 11) Determine the x-intercept for the function: $f(x) = \frac{1}{2}x - 7$

A) -7

B) $\frac{1}{2}$

C) 7

D) 14

_____ 12) Determine the range given the domain $\{-3, 0, \frac{1}{2}\}$ for the function: $h(x) = \frac{1}{2}x + 1$.

A) $\{\frac{1}{2}, 1, \frac{5}{4}\}$

B) $\{-\frac{1}{2}, 1, \frac{5}{4}\}$

C) $\{-\frac{1}{2}, 1, \frac{4}{5}\}$

D) $\{\frac{1}{2}, 1, 5\}$

Lesson 10
Writing Linear Equations

Outline

Part A:
Writing Equations in Slope-Intercept Form

- Slope-Intercept Form Given Slope and a Point
- Slope-Intercept Form Given Two Points

Part B:
Applications of Linear Equations

- Applications of Linear Equations

Targeted Review

Vocabulary

- y-intercept
- extraneous information

10A EXPLORE

Part A: Writing Equations in Slope-Intercept Form

Objectives

In this part of the lesson, you will learn about writing equations in slope-intercept form.

By the end of this lesson you will be able to do the following:

- ✓ Write an equation in slope-intercept form given the slope and one point.
- ✓ Write an equation in slope-intercept form given two points.

Why?

As you learned in Lesson 9, one of the most important things you will do in Algebra 1 is use the slope formula to solve linear equations. In this lesson, you will continue to master using slope.

Warm Up

Use your formula sheet to answer the following questions.

1) What is the formula for slope?

2) What is point-slope form?

3) What is slope-intercept form?

▶ Slope-Intercept Form Given Slope and a Point

- Point-slope form and slope-intercept form are similar in that they both use _____ and _____.

- Point-slope form can be written using any _____ from the graph.

- Slope-intercept form uses slope and a specific point, the _____.

- Slope-intercept form can be derived from a point on the line and the slope of the line by writing an equation for the line in _____ first, and then _____.

EXPLORE 10A

Example 1

The linear function $g(x)$ has a slope of $\frac{3}{4}$ and passes through the point $(6, 4)$.

Write the equation for $g(x)$ in slope-intercept form.

Plan Identify the information from the problem. $m = \frac{3}{4}, (6, 4)$
Write an equation in point-slope form using the given slope and point. Then solve for y.

Implement   **Explain**

◂ Distribute $\frac{3}{4}$ on the right side of the equation

◂ Isolate y

◂ Combine like terms

◂ Simplify fractions

◂ Rewrite in function notation

> Now that you know multiple forms for equations of lines, always read carefully so that your answer is in the correct form.

☑ Checkpoint

Write the equation in slope-intercept form.

Point: $(10, 5)$ Slope: $\frac{4}{5}$

10A EXPLORE

▶ Slope-Intercept Form Given Two Points

- Slope-intercept form of a line can be written if _____ for the function are known.

- Because _____,
either known point can be selected to write the equation in slope-intercept form.

- Picking the point with the fewest number of _____ values is a good idea so there is less of a chance of making a mistake with the signs of the numbers.

Example 2

Write the equation of a line in slope-intercept form that passes through the points (3, 4) and (2, 7).

Plan Find the slope.
Pick one point and substitute it into point-slope form.
Isolate y.

Implement

Slope: $m = \frac{\Delta y}{\Delta x} =$

Solve for y using (2, 7):

$y - 7 = -3(x - 2)$

$y - 7 = -3x + 6$

$y = -3x + 13$

Solve for y using (3, 4):

Example 3

Use the equation $y = \frac{1}{4}x + 11$ to:

A) Find the value of x when $y = 7$.

B) Find the value of y when $x = 8$.

Plan Write the information that you know as an ordered pair.
Then substitute the information into the linear equation and solve.

Implement

A) $(x, 7)$

$7 = \frac{1}{4}x + 11$

$-4 = \frac{1}{4}x$

$x = -16$

$(-16, 7)$

B) $(8, y)$

☑ Checkpoint

Write an equation of a line in slope-intercept form that passes through both $(4, -4)$ and $(-1, 6)$.

10A   PRACTICE 1

✏️ Practice 1

Complete the problems on a separate sheet of paper.

All equations in Lesson 10 should be written in slope-intercept form unless stated otherwise.

Write the equation of the line.

1) $y + 9 = \frac{1}{2}(x - 3)$

2) $y - 1 = -\frac{5}{3}(x + 2)$

3) $y - 2 = 4(x + 3)$

4) $y + \frac{3}{4} = 12\left(x - \frac{1}{4}\right)$

Write the equation of the line in slope-intercept form given the slope and a point on the line.

5) $m = -\frac{1}{2}$ and $(2, 5)$

6) $m = 3$ and $(-1, 0)$

7) $m = -\frac{4}{5}$ and $(10, 7)$

Reanne was walking at a rate of 4 miles per hour. One hour later, she was at mile marker 7.

8) What is the rate of change? Explain.

9) What is the ordered pair, *(time, distance)*, that is provided in the information?

10) Write an equation that represents Reanne's walk.

11) At what mile marker will she be after two and a half hours?

An internet company charges $42 per month plus a one-time setup fee. After five months, Joey had paid $259.50.

12) What is the rate of change?

13) Define the variables as an ordered pair in words.

14) What is the ordered pair in the given information?

15) Write an equation that represents this scenario.

16) Joey noticed that after having internet for *x* months, he had paid a total of $511.50. How many months did Joey have the internet? Show your work.

Write an equation for the line that passes through the given points.

17) $(-3, 7)$ and $(4, 14)$

   A) Use the slope formula to determine the slope.

   B) Use the point $(-3, 7)$ to write the point slope form.

   C) Write the equation in slope-intercept form.

Write an equation for the line that passes through the given points.

18) $(-2, -6)$ and $(-4, 3)$

19) $(-5, 2)$ and $(1, 4)$

20) $(-7, 9)$ and $(-4, 7)$

---

21) Write an equation for the given functions using the graph provided.

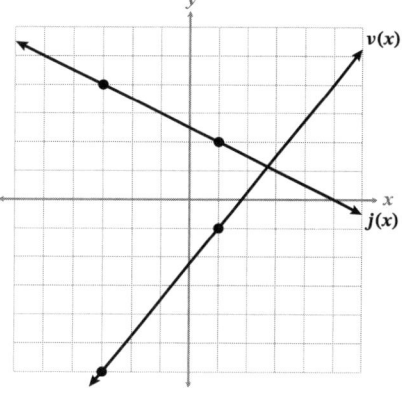

10A MASTERY CHECK

Mastery Check

✎ Show What You Know

Use the ordered pairs (−1, 1) and (2, 3).

A) Graph the ordered pairs and draw the line formed between these points.

B) Find the equation of the line algebraically.

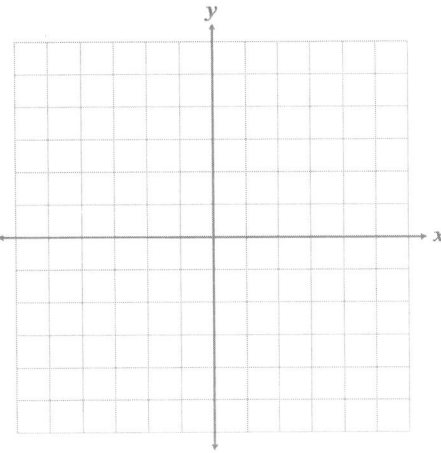

C) Explain why it is not possible to determine either intercept from the graph accurately.

D) Solve for the x-intercept using your equation.

E) How has this line been translated related to the origin? Explain.

🔊 Say What You Know

In your own words, talk about what you have learned using the objectives for this part of the lesson and your work on this page.

Practice 2

Complete the problems on a separate sheet of paper.

All equations in Lesson 10 should be written in slope-intercept form unless stated otherwise.

Write the equation of the line.

1) $y - 2 = \frac{6}{5}\left(x - \frac{1}{2}\right)$

2) $y - 1 = -\frac{1}{3}(x - 5)$

3) $y + 3 = \frac{3}{2}(x + 1)$

4) $y - 7 = -\frac{3}{4}(x + 5)$

Write an equation of the line given the slope and a point on the line.

5) $m = -7; \left(-\frac{1}{2}, 0\right)$

6) $m = \frac{5}{7}; (-14, 2)$

7) $m = \frac{2}{3}; (3, 2)$

The price of fuel at a local station was $3.25 per gallon. Katrina used a gift card to put 2 gallons in her car. When she saw the receipt for her purchase, there was still $20 left on the gift card.

8) What is the ordered pair, *(gallons bought, money remaining)*, provided in the information?

9) What is the slope? Why is it negative?

10) Write the linear equation for this scenario.

11) How many gallons of fuel could Katrina have bought using the entire gift card? Round to the nearest hundredth.

Write the equation for a line given two points from the line.

12) $(1, 3)$ and $(-1, -5)$

13) $(-4, -1)$ and $(-2, 2)$

14) $(-3, 7)$ and $(6, -5)$

---

15) Use the marked points to find the equation in slope-intercept form. Write your final answer in function notation.

Find the equation of each line that forms triangle CDE in slope-intercept form. List the ordered pairs you are using before solving algebraically.

16) Line $\overleftrightarrow{CD}$

17) Line $\overleftrightarrow{DE}$

18) Line $\overleftrightarrow{CE}$

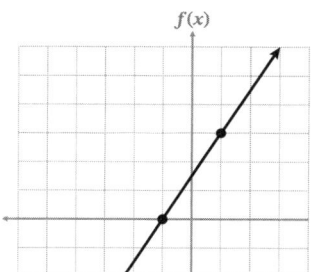

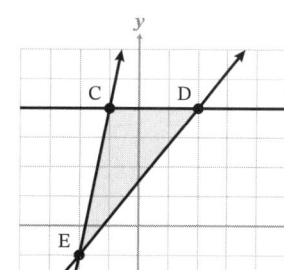

Part B: Applications of Linear Equations

Objectives

In this part of the lesson, you will learn about applications of linear equations.

By the end of this lesson, you will be able to do the following:

- Write an equation in slope-intercept form given any type of scenario.
- Explain what a given point, the slope, and the x- and y-intercept represent within the context of a word problem.

Why?

It's not enough to simply know how to solve linear equations. For them to be useful, you need to understand how to apply them properly to real-world scenarios.

Warm Up

1) What does the word "per" help determine in a word problem?

2) When graphing an equation in slope-intercept form, where do you start the graph?

Applications of Linear Equations

- Your advantage over a calculator when solving linear equations is that you understand _____ of slope, the y-intercept, and all points.

- Once you have a linear equation, you should be able to explain each part as it relates to a given problem. This means you will be able to explain the following:

    - Slope as a _____.

    - The _____ as more than just a starting point on the graph.

    - Any _____ as it relates to the equation and the context of the problem.

- _____ information is information that is irrelevant to the problem and is sometimes provided to distract from the needed values.

- Helpful ways to plan when given an application problem (word problem):

    - Identify _____ by highlighting, circling, or underlining it.

    - Label _____.

    - Define the _____ as an ordered pair.

    - Use your _____ to write down the formulas you will use to solve the problem.

Example 1

Carrie got in her car at 8 a.m. to drive to work. After a half-hour, she had traveled 25 miles. When she parked her car 45 minutes after leaving her house, she had traveled 37.5 miles.

Write the linear equation that represents the scenario and explain what the slope and y-intercept represent in context.

Plan Find the key information, then circle, highlight, or underline it.
Write the ordered pair in words.
Organize the four values provided.
Find the slope and explain its meaning.
Write the linear equation.
Find the y-intercept and explain its meaning.

Implement

Ordered pair in words:

Values provided:

Slope: $m = \frac{37.5 - 25}{0.75 - 0.5} = \frac{12.5}{0.25} = \frac{50}{1}$

Explain

◂ When $m = \frac{50}{1}$, Carrie was traveling 50 miles every hour, or Carrie was driving 50 mph.

Linear equation (use either point):
$m = 50, (0.5, 25)$

$y - 25 = 50(x - 0.5)$

$y - 25 = 50x - 25$

$y = 50x$ ◂

Determine the time that Carrie got to work.

8:00 a.m. + 45 min =

10B EXPLORE

Example 2

Marco is a residential electrician. He charges an assessment fee for each job plus $22 per hour of work completed, rounded up to the quarter-hour. Marco's last job took 3.75 hours, and he charged the customer $127.50.

Write an equation that Marco can use to find his earnings for a job of any length of time.

Plan Identify the key information.
Write the ordered pair in words and organize your values.
Write the equation and explain its meaning.

Implement

Ordered pair in words: Equation:

Values provided:

Explain

If Marco needed to earn at least $160 to pay his bills this week, how many hours would you suggest he work?

EXPLORE 10B

☑ Checkpoint

JaJuan opened a savings account and plans to deposit the same amount of money in the account weekly. JaJuan had $250 three weeks after opening his savings account. Seven weeks after opening the account, he had saved $550.

A) Write an equation to find his savings for any number of weeks.

B) Name the rate of change. Explain what it means in this problem.

C) Name the y-intercept. Explain what it means in this problem.

D) If JaJuan is saving to purchase a used car that costs $3,500, how many weeks will he need to save?

Practice 1

Complete the problems on a separate sheet of paper.

Write an equation in slope-intercept form. Make sure to label information and watch for extraneous information.

1) Padma started a vegetable garden in the spring. Once the plants sprouted, she began measuring them daily to track the plant growth. Padma noticed that her plants grew 2 cm per day, and on day 13, she measured the plants at 27 cm. Write an equation to show plant growth for any number of days.

2) Marshall eats eggs every morning for breakfast. He went to the store in the afternoon of August 31st and purchased 5 dozen eggs (60 eggs) for September. On September 25th, 10 eggs were remaining. Write an equation to represent Marshall's egg consumption for September.

Cedar-Oak School was selling yearbooks. The school was charged a publishing fee of $325 and a $35 per book fee for the yearbooks.

3) Write an equation to represent the cost for the yearbooks for any number of books that the school will purchase.

4) How much will it cost for the school to purchase 100 yearbooks?

5) A propane company fills a 100-gallon tank to 85% capacity. Edward had propane delivered from the company, and after one month, the tank was 80% full. Four months later, Edward checked the tank and saw it was now 65% full. Write an equation to demonstrate propane usage if the same amount is used each month.

A plane left an airport in Philadelphia at 9 a.m. (EST) on a direct flight to Los Angeles. At 2:30 p.m. (EST), the plane landed after flying 2,400 miles. (EST = Eastern Standard Time).

6) Write an equation for the flight from Philadelphia to Los Angeles. Round to the nearest whole number.

7) What was the cruising speed of the plane?

Tarek was installing new drywall in his house. He recorded his costs for drywall in the table.

Sheets of drywall	Cost ($)
5	219.90
7	247.86
11	303.78
15	359.70

8) Write an equation to calculate his cost for any number of sheets of drywall and supplies.

9) Name the y-intercept. Explain what you think it means in this problem. Why is it greater than 0?

10) Explain what the point (11, 303.78) represents.

11) Suppose Tarek only wanted the cost of drywall. What would he pay for only three sheets of drywall?

Garrison paid a one-time fee for a music card to purchase any number of songs on the site for the same price. After purchasing three songs, he had spent $21 for the card and the songs. After purchasing six songs, he paid $30 for the card and the songs.

12) Write an equation for the given scenario in slope-intercept form, where n is the number of songs and d is dollars spent.

13) How many songs could Garrison purchase along with the music card for $45?

14) What does the slope represent for this function?

15) What does the y-intercept represent for this function?

Kristina was emptying 10,000 gallons of water from her swimming pool. The water was *removed* from the pool at a rate of 125 gallons per minute.

16) Write an equation for the given scenario in slope-intercept form.

17) What does the slope represent for this function? What does the y-intercept represent for this function?

18) How long would it take to empty the pool? How do you know this?

The graph represents the company T-shirts R Cool. The y-intercept is the amount of money borrowed from the bank to start the company.

19) Find the equation of the line.

20) Explain the meaning of the marked point $(10, -100)$.

21) A **break-even point** is a business term used when you have earned back all of the money you invested into your company but have not earned any income (money) yet. Find the x-intercept (or break-even point) using your equation. Explain what the x-intercept means.

22) Why do you think the points on the line are marked?

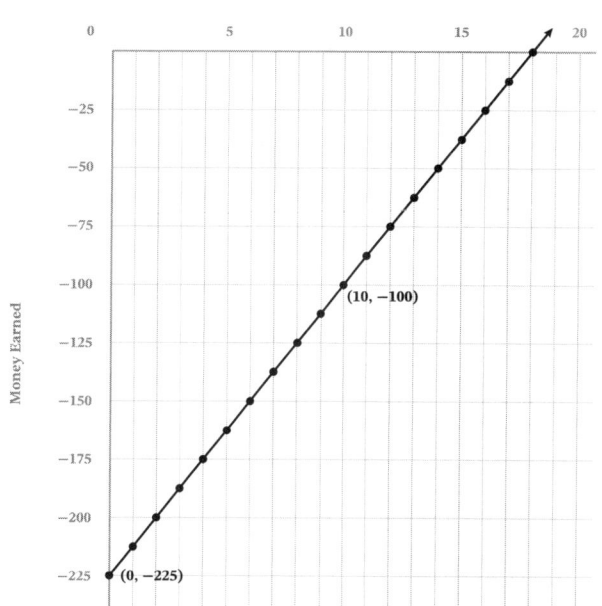

10B MASTERY CHECK

✓ Mastery Check

✎ Show What You Know

Three siblings—Mark, Steven, and Elizabeth—all completed a 5K charity race and raised a combined total of $1,200. The graph shows how each person did in the race.

A) Find the equation, in function notation, for each sibling.

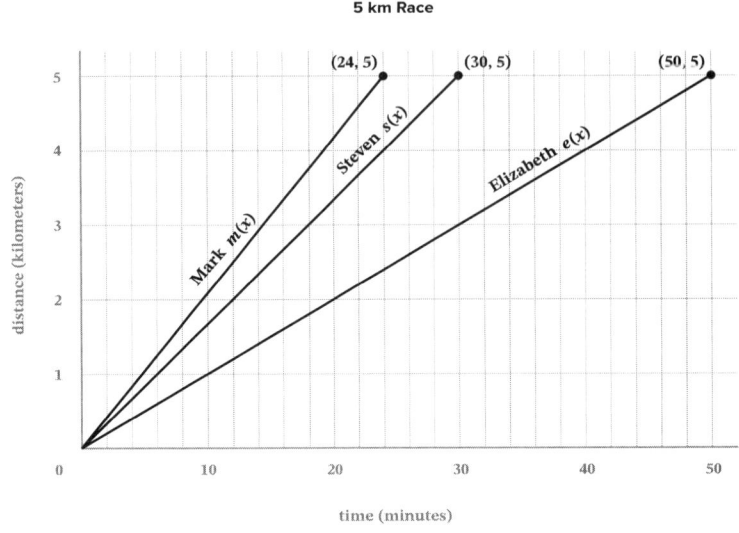

B) Explain the rate of change for each runner.

C) Who won the race? Justify your response with mathematical details.

D) Why do all of the graphs stop when $y = 5$?

·ılıl· Say What You Know

In your own words, talk about what you have learned using the objectives for this part of the lesson and your work on this page.

Practice 2

Complete the problems on a separate sheet of paper.

Write the equations in slope-intercept form. Watch for extraneous information.

Miguel needed to dig 300 feet to install a well for his new house. The construction manager said that this would take 40 hours to complete. Miguel hopes that the digging will only take 30 hours.

1) List the ordered pairs for the problem and explain their meaning.

2) Write an equation to find the rate the drill is digging at.

A class trip was being planned to the science museum. The school would pay the transportation fee, and the students would pay the cost of admission. Mrs. Wallace created a table to track total expenses.

3) Write an equation representing the class trip.

4) How much did the bus ride cost for the trip? Explain how you know this.

Students	Cost ($)
15	550
30	700
45	1,250
60	14,000

Devon had a bank account from which he withdrew $55 each week for his bills. There was $3,170 remaining in the account after six weeks.

5) Write an equation to represent Devon's spending.

6) At what week will he run out of money?

Agnes was getting her swimming pool filled by the fire company and predicted it would take them 2.5 hours to complete. The fire chief said that the fire hose has a flow rate of 150 gallons per minute. After 20 minutes, the pool had 2,200 gallons of water in it.

7) Write an equation to model the pool being filled.

8) If it took 86 minutes to fill the pool, how much water was in the pool when it was filled?

The graph shows a Bike Ride over the course of several hours. The distance is in miles.

9) Find the equation of the line for part A.

10) What does the rate of change mean for part A?

11) Find the slope for part C.

12) Find the slope for part D.

13) During which part of the graph did the rider travel the farthest? Justify your response with mathematical details.

14) From the graph, what type of line segments are parts B and E? What do parts B and E represent in the context of the problem?

15) What does the intercept (0, 0) mean for this scenario?

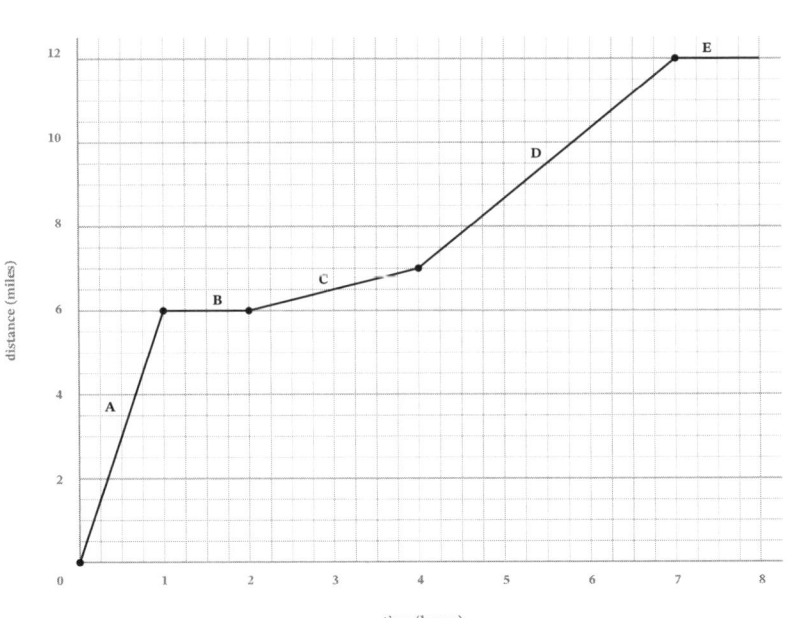

A team of students will be taking an overnight trip. The coach needs to pay a one-night deposit to reserve rooms in a hotel and purchase a parking space. The deposit for six rooms and the parking space will cost $780. The deposit for twelve rooms and the parking space will cost $1,500. Write an equation to provide the deposit, $d(r)$, needed to reserve the appropriate number of rooms and the parking space.

16) Write an equation for the given scenario in slope-intercept form and function notation.

17) How much would it cost to reserve 15 rooms and the parking space?

18) What does the slope represent for this function?

19) What does the y-intercept represent for this function?

A full 500-gallon propane tank started leaking at a rate of $2\frac{1}{2}$ gallons per minute.

20) Write an equation for the given scenario in slope-intercept form.

21) What does the slope represent for this function? What does the y-intercept represent for this function?

22) How long would it take for the tank to be empty?

Christina used 2.9 GB of data and owed $75 for her cell phone bill in January. In February, she doubled her data, and her bill was $75.

23) Write an equation to represent her cell phone bill each month.

24) What type of cell phone plan do you think Christina has (unlimited data or limited data)? Explain.

25) What information was extraneous in the problem?

TARGETED REVIEW 10

Targeted Review

In the Targeted Review, you will practice topics you have mastered in earlier lessons. Reviewing these concepts will help you be successful as you work through this unit.

Complete the problems on a separate sheet of paper.

1) Name the x and y-intercepts as ordered pairs from the table.

2) Name the domain and the range for the table.

3) The table has a constant rate of change. Find the slope using any two points.

x	$f(x)$
-3	$-\frac{3}{2}$
0	$-\frac{1}{2}$
$\frac{3}{2}$	0
3	$\frac{1}{2}$

4) Mark the x and y-intercepts on the graph. Write the intercepts as ordered pairs.

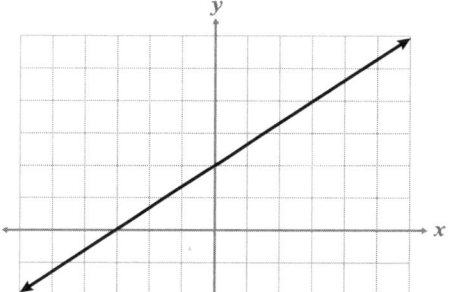

5) Write an equation in point-slope form. $m = -\frac{5}{2}, (-1, 15)$

6) Loretta mailed two cards (c) each day (d). Determine which variable is dependent and independent for the scenario and write an equation in function notation.

Use the graph to answer the following questions.

7) Does $w(2) = 1$? Explain.

8) Does $w(3) = 0$? Explain.

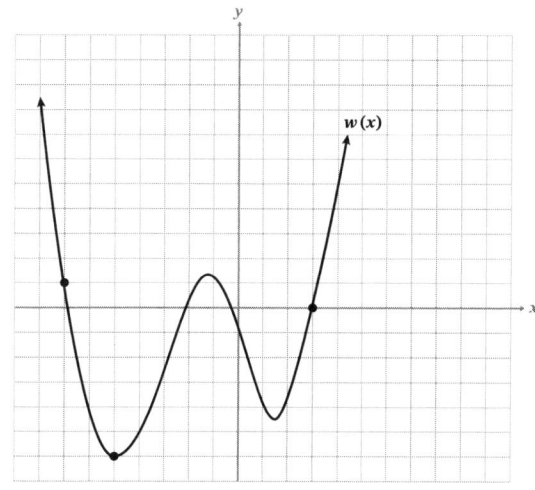

TARGETED REVIEW 10

Solve. Graph the solution(s) on a number line.

9) $|-2x - 6| = 3$

10) $|x - 2| - 4 > -4$

11) How is graphing on a number line different from the graphing in Unit 2?

Multiple Choice

_____ 12) Find the value of r.

$m = 4, (-3, 2)$ and $(r, 12)$

      A) -2

      B) $-\frac{1}{2}$

      C) $\frac{1}{2}$

      D)   37

_____ 13) Determine the equation in slope-intercept form when

$m = 6$ and $b = \frac{1}{3}$

      A) $y = \frac{1}{3}x + 6$

      B) $y - \frac{1}{3} = 6(x - 0)$

      C) $y = 6x + \frac{1}{3}$

      D) $y - 6 = \frac{1}{3}(x - 0)$

Lesson 11
More Forms of Lines

Outline

Part A
Standard Form

- Graphing an Equation in Standard Form
- Writing an Equation in Standard Form
- Comparing Forms of Linear Equations

Part B
Horizontal and Vertical Lines

- Horizontal Lines
- Vertical Lines

Targeted Review

Vocabulary

- standard form
- horizontal line
- vertical line

Part A: Standard Form

Objectives

In this part of the lesson, you will learn about standard form.

By the end of this lesson you will be able to do the following:

- Solve for the x- and y-intercepts from standard form and use the intercepts to create a graph of the line.
- Convert the equation of a line to standard form and determine the slope and intercept formulas found in a linear equation in standard form.

Why?

Until now, you have primarily worked with lines in point-slope and slope-intercept form. It is also important to understand how standard form is used.

Warm Up

Write the equations in slope-intercept form.

1) $5x + 6y = 0$

2) $2x - y = 4$

Graphing an Equation in Standard Form

- $Ax + By = C$ is a linear equation in _____ form.

- For equations in standard form, the following must be true:

    - A must be a _____ number.

    - B and C must be _____.

    - A, B, and C have a _____ of 1.

- To graph a line from standard form, find the _____

    because it takes two points to draw a line.

- The x-intercept is always in the form _____.

- The y-intercept is always in the form _____.

Example 1

Graph the equation of the line: $3x - 2y = 6$

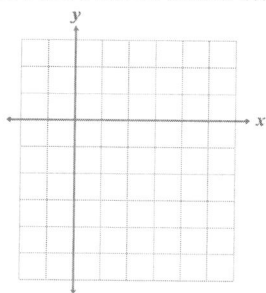

Plan   Solve for the *x*-intercept.
Solve for the *y*-intercept.
Write answers as ordered pairs.
Graph the points.
Draw the line through the two points.

Implement

Solve for the *x*-intercept $(a, 0)$.   Solve for the *y*-intercept $(0, b)$.

$3x - 2y = 6$

$3a - 2(0) = 6$

$3a - 0 = 6$

$3a = 6$

$a = 2$

The *x*-intercept is 2, and the point is $(2, 0)$.

> In some instances, you may prefer to use slope-intercept form rather than standard form when graphing. This would especially be true if $C = 0$.

Example 2

Graph the equation of the line: $5x + 6y = 30$

x-intercept   *y*-intercept

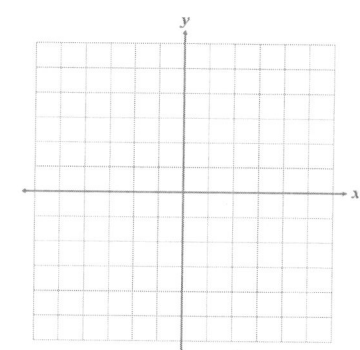

11A EXPLORE

✓ Checkpoint

Graph from standard form. Mark and label your x and y-intercept on the graph.

$5x + 4y = -20$

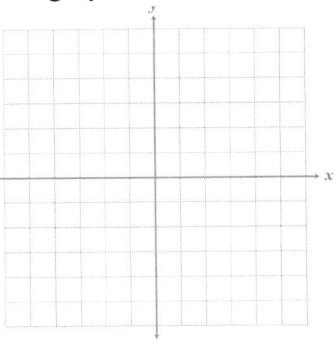

▶ Writing an Equation in Standard Form

■ The standard form of a line creates a single solution for each line, which allows different lines to be _____.

Example 3

Convert the equation to standard form: $y = \frac{3}{4}x - 2$

Plan Move both variables to the same side of the equation.
Use the rules to simplify until the equation is in standard form.

Implement   **Explain**

$$y = \frac{3}{4}x - 2$$

$$-\frac{3}{4}x \quad -\frac{3}{4}x$$ ◀ Addition Property of Equality

$$-\frac{3}{4}x + y = -2$$

$$(-4)\left(-\frac{3}{4}x + y\right) = (-4)(-2)$$ ◀ Multiplication Property of Equality.

$$3x - 4y = 8$$ ◀ GCF $(3, -4, 8) = 1$, this line is in standard form.

EXPLORE 11A

Example 4

Convert the equation given in point-slope form to standard form: $y - 1 = \frac{3}{5}(x + 4)$

Plan   Solve for y.
Move both variables to the same side of the equation.
Simplify the equation to standard form.

Implement

$$y - 1 = \frac{3}{5}(x + 4)$$

Explain

◀ Move both variables to the same side of the equation

◀ Simplify the equation

☑ Checkpoint

Write the following equation in standard form: $y - 4 = \frac{2}{3}(x - 9)$

▶ Comparing Forms of Linear Equations

- Knowing the formula for slope and y-intercept can save you from _____ all the equations in the _____ form.

11A EXPLORE

Example 5

Gary found the following list of equations. However, he needed to know the slope for each line represented on the list.

Find the formula that would help Gary find the slope for each line.

A)	$Ax + By = C$
B)	$2x + 3y = 6$
C)	$4x - 2y = 7$
D)	$-\frac{2}{3}x + \frac{1}{4}y = \frac{3}{4}$

Plan First, y is multiplied by B.
Then Ax is added to y.

A) **Implement**   **Explain**

$Ax + By = C$ ◂ Given

$By = -Ax + C$ ◂ Addition Property of Equality

$\left(\frac{1}{B}\right)By = \left(\frac{1}{B}\right)(-Ax + C)$ ◂ Multiplication Property of Equality

$y = -\left(\frac{A}{B}\right)x + \frac{C}{B}$ ◂ Distributive Property

This is the general rule, or formula, for finding the slope and y-intercept when given an equation in standard form.

slope: y-intercept:

Use the formula to find the slope for the remaining equations in this example.
First identify A, B, and C.

B) $2x + 3y = 6$
$A = 2 \quad B = 3 \quad C = 6$
slope: $m = -\left(\frac{A}{B}\right) = -\frac{2}{3} = -\frac{2}{3}$

C) $4x - 2y = 7$

D) $-\frac{2}{3}x + \frac{1}{4}y = \frac{3}{4}$

☑ **Checkpoint**

Given $5x + 3y = 2$, find the slope and the y-intercept using the formula.

Practice 1

Complete the practice problems on a separate sheet of paper.

Find the x-intercept and y-intercept for each equation. Graph the line on a coordinate plane.

1) $3x - y = 6$
2) $x - 3y = 6$

Find the x-intercept and y-intercept for each equation.

3) $5x + 2y = 3$
4) $3x - y = -19$
5) $5x - 6y = 72$

6) Jenna planned on making sandwiches for lunch this week. She had $10 that she planned to spend on sliced turkey or cheese. Turkey is $10 per pound, and cheese is $5 per pound. How much would Jenna be able to purchase if she only purchased turkey or only purchased cheese using the equation $10x + 5y = 10$ (x: turkey, y: cheese)?

Write an equation in standard form. Identify A, B, C.

7) $y - 6 = \frac{2}{7}(x - 7)$
8) $y = -\frac{1}{9}x + 11$

Identify A, B, C in the given equation. Find the slope and y-intercept from standard form.

3) $x - 2y = 4$
4) $8x - 5y = -6$

11) Given the equations, for a, b, and c, find the slope and y-intercept.

12) Which equation has the smallest slope? Explain.

13) Which equations have equal y-intercepts?

| Line a: $2x + 3y = 6$ |
| Line b: $y = \frac{2}{3}x + 3$ |
| Line c: $3x + 2y = 6$ |

14) Given the equations, for p, q, and r, find the slope and y-intercept.

15) List the lines in order from greatest to least slope.

| Line p: $x - y = P$ |
| Line q: $4x + 7y = Q$ |
| Line r: $6x - 5y = R$ |

16) Identify the slope and y-intercept for: $x + Ey = F$

17) What formula do you use to determine the y-intercept from standard form?

18) What formula do you use to determine the slope from standard form?

11A MASTERY CHECK

Mastery Check

Show What You Know

Austin was making sandwiches for a party. He used the ordered pair (cheese, ham). Austin can purchase 6 pounds of ham with 0 pounds of cheese. Likewise, he can purchase 10 pounds of cheese and 0 pounds of ham, which he determined using $y = -\frac{3}{5}x + 6$.

A) Write the equation $y = -\frac{3}{5}x + 6$ in standard form.

B) The numbers in the equation from part A represent dollar amounts.

How much does one pound of cheese cost?

How much does one pound of ham cost?

What is the total amount of money Austin has to spend?

C) Austin decided to increase his budget to $90 to ensure that everyone would have enough to eat. Write the new equation in standard form and find the intercepts as ordered pairs.

D) Graph your equation from Part C.

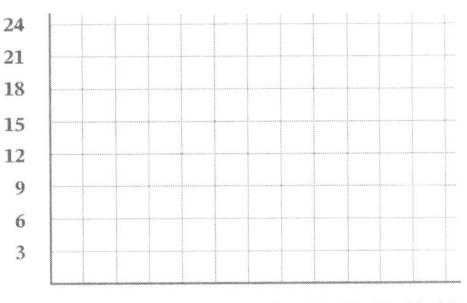

E) Would Austin be able to purchase 15 pounds of cheese and 9 pounds of ham? 18 pounds of cheese and 9 pounds of ham? Explain your thinking for each purchase.

Say What You Know

In your own words, talk about what you have learned using the objectives for this part of the lesson and your work on this page.

Practice 2

Complete the practice problems on a separate sheet of paper.

Find the *x*-intercept and *y*-intercept. Graph the line on a coordinate plane.

1) $2x - y = 8$
2) $5x + 3y = -15$

Find the *x*-intercept and *y*-intercept for each equation.

3) $2x + 5y = 10$
4) $12x - 5y = 30$
5) $3x + 4y = 12$

6) While attending a sporting event, a family was deciding between hot dogs for $5 and hamburgers for $8. They had a spending budget of $40. Using (*x*: hot dogs, *y*: hamburgers), find how many hot dogs or hamburgers could be purchased using the equation $5x + 8y = 40$.

Write the equation in standard form. Then identify the slope and *y*-intercept.

7) $y = \frac{1}{3}(x + 6) - 1$
8) $\frac{x}{2} + \frac{y}{3} = 1$
9) $y - \frac{1}{5} = \frac{3}{7}(x + 1)$
10) $y = \frac{2}{3}\left(x - \frac{1}{2}\right) + 4$

Use the table to compare the equations.

11) List the equations from least to greatest *y*-intercept.

12) List the equations from greatest to least slope.

Line *a*:	$y - 1 = -2(x - 1)$
Line *b*:	$2x - y = 2$
Line *c*:	$y = -\frac{4}{3}x + 4$
Line *d*:	$4x - 3y = -3$

Given: $y = -Ex + H$

13) Determine the slope (*m*) and *y*-intercept (*b*).

14) Write the equation in standard form.

15) Use the results in problems 13 and 14 to identify the value of *E* and *H* in $15x + y = 30$, and determine the slope and *y*-intercept.

11B EXPLORE

Part B: Horizontal and Vertical Lines

Objectives

In this part of the lesson, you will learn about horizontal and vertical lines.

By the end of this lesson, you will be able to do the following:

- Graph horizontal and vertical lines.
- Find the domain, range, slope, and y-intercept for horizontal and vertical lines.
- Determine the equation of horizontal and vertical lines that pass through a given point.

Why?

Horizontal and vertical lines do not show up often in algebra. However, it is critical to understand their uniqueness when you encounter problems that include them.

Warm Up

Use the slope formula to find the slope between two points.

1) $(2, 11)$ and $(2, 6)$

2) $(24, 7)$ and $(11, 7)$

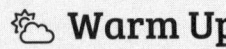 Horizontal Lines

- A horizontal line (↔) is a line in which the input (*x*-values) can be _____, and the output (*y*-value) is equal to the _____.

- Because the graph of a horizontal line passes the vertical line test (VLT), it is a _____.

- The equation for a horizontal line is $y = 0x + b$ or more commonly, _____.

- When you see that *y* equals a number, you should visualize a _____ line.

- The slope of _____ horizontal line is _____.

- The _____ for a linear function is all real numbers.

- Remember to check the _____ of each axis when reading a graph.

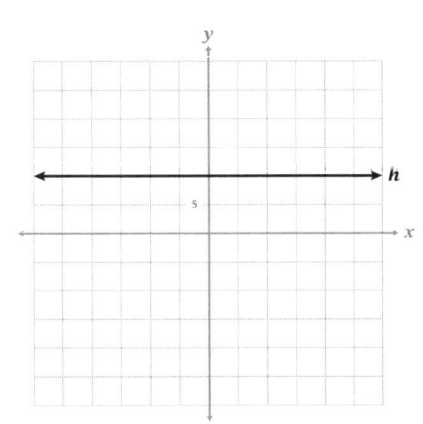

EXPLORE 11B

Example 1

Write the equation for line *g* shown in the graph. Name the domain and range.

Plan Find the scale of the coordinate plane.
Determine the slope and *y*-intercept.
Write the equation in slope-intercept form for this line.
Determine the domain and range.

Implement

x-axis scale: 2

y-axis scale: 3

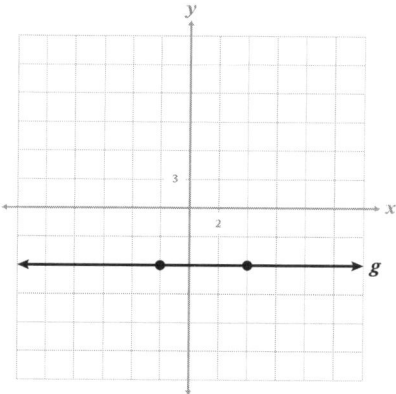

Domain: {ℝ}   This will always be the domain for a horizontal line (and linear functions).
Range: {−6}   The range will always be a single number since the output remains the same.

Example 2

Write the equation of the horizontal line that passes through the point (−2, 4).

Plan Graph the point on a plane.
Draw a horizontal line through the point.
Write the equation in slope-intercept form.

Implement

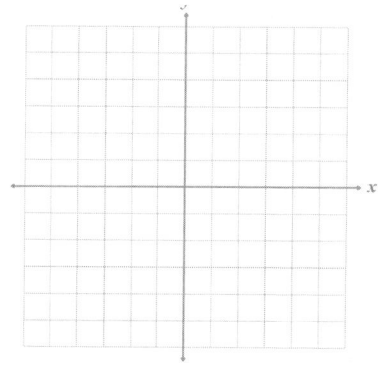

- Notice that the equation is the same as the *y*-coordinate of _____ included on the line.

- The _____ of the ordered pair will also be the number you use for the _____ of a horizontal line.

☑ Checkpoint

Given the graph, find the domain and range of the horizontal line. Write the equation of the line.

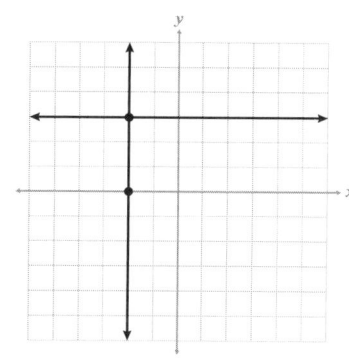

11B EXPLORE

▶ Vertical Lines

- A vertical line (↕) is a line that has a single _____, and the _____ can be any real number.

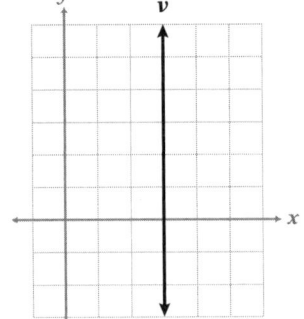

- The graph of a vertical line fails the vertical line test (VLT) because there is one _____ and the _____ is _____.

- Because it fails the vertical line test (VLT), a vertical line is not a _____.

- Writing the equation for a vertical line in _____ form is not possible because $y = mx + b$ becomes $y =$ (undefined)$x +$ (does not exist).

- The equation for a vertical line is _____, where a is the x-intercept.

Example 3

Write the equation for line c shown in the graph. Name the domain and range.

Plan   Find the scale of the coordinate plane.
Determine the slope and y-intercept.
Write the slope-intercept form of the equation of the line.
Determine domain and range.

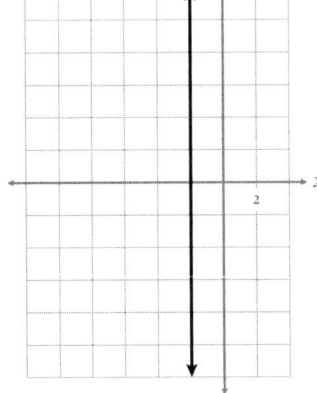

Implement

x-axis scale: 2 $m =$

y-axis scale: 1 $b =$

x-intercept:

The equation for this line:

Domain:

Range:

Example 4

Write the equation of the vertical line that passes through the point (−2, 4).

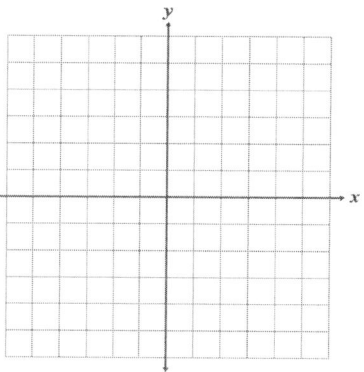

- The equation is the _____ of any point that is on the line.

- Vertical lines do not need to be graphed to find the equation because the formula _____ is known.

☑ Checkpoint

Provide the equation of the vertical line and the horizontal line that passes through the point (−7, 3). Name the slope for each line.

Vertical line: Horizontal line:

11B PRACTICE 1

✏️ Practice 1

Complete the practice problems on a separate sheet of paper.

Using the graph below, answer the following questions.

1) What is the slope of line h?
2) What is the y-intercept for line h?
3) Write the equation for line h.
4) Does line h represent a function? Explain.

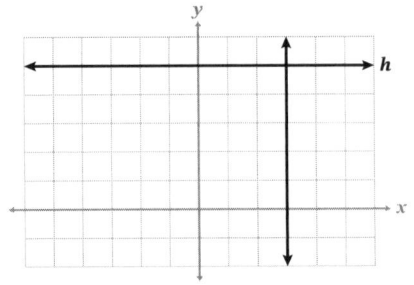

5) Find the slope of line f that passes through the points $(1, -2)$ and $(-5, -2)$.
6) Write the equation of line f.
7) Does line f also pass through the point $(10, -2)$? Explain.
8) Is the line horizontal or vertical? Without graphing, explain your reasoning.

9) Graph the line $y = 15$ on a coordinate plane.
10) Graph the horizontal line passing through the point $(4, -6)$. Find the equation of the horizontal line passing through the point $(4, -6)$. Name the domain and range.

11) Find the domain and range for the horizontal line $y = -9$.
12) Nicole wrote the equation for the horizontal line that passes through the point $(-2, -6)$ as $y = -2$. Explain her mistake, and provide the correct equation.

Using the graph of line v, answer the following questions:

13) What is the slope of line v?
14) What is the y-intercept for line v?
15) Write the equation for line v.
16) Does line v represent a function? Explain.

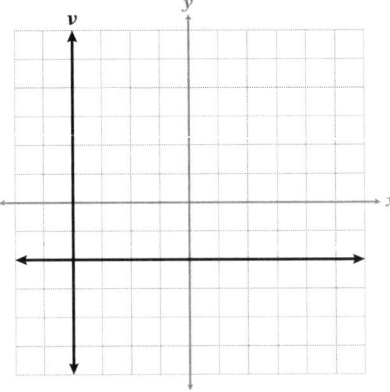

17) Find the slope of the line that passes through the points $(-7, 1)$ and $(-7, 13)$.

18) Write the equation of the line that passes through the points $(-7, 1)$ and $(-7, 13)$.

19) Will this line also pass through the point $(1, 1)$? Explain.

20) Is the line horizontal or vertical? Without graphing, explain your reasoning.

---

21) Graph the line $x = 8$ on a coordinate plane.

22) Graph the vertical line and the horizontal line passing through the point $(3, 1)$.

Find the equation of the vertical line passing through the point $(3, 1)$. Name the domain and range.

23) Find the domain and range for the vertical line $x = -1$.

24) Determine if the point $(10, -7)$ is on the line $x = 10$. Why or why not? Name the vertical and horizontal lines going through the given point.

11B MASTERY CHECK

Mastery Check

Show What You Know

A group of students was given the point $(-5, 1)$. Each student was asked to state a fact about the point.

A) Jeremiah said the horizontal line passing through this point is $x = -5$, and Matthew said the horizontal line passing through this point is $y = 1$. Who is correct? Explain.

B) Graph the ordered pairs and connect the points to form rectangle $PQRS$.

$P(-3, 4), Q(5, 4), R(5, -2), S(-3, -2)$

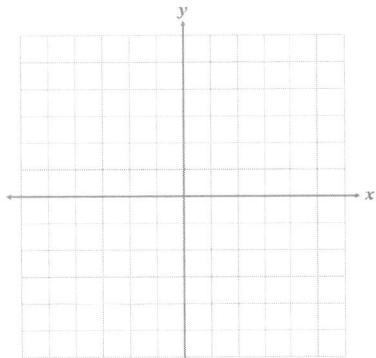

C) Write the equation for each side of rectangle $PQRS$.

$\overleftrightarrow{PQ}$:

$\overleftrightarrow{QR}$:

$\overleftrightarrow{RS}$:

$\overleftrightarrow{PS}$:

D) Jeremiah decided to move point P on the graph to $(-4, -3)$. Without doing any calculations, explain how your equations in part C will change.

Say What You Know

In your own words, talk about what you have learned using the objectives for this part of the lesson and your work on this page.

Practice 2

Complete the practice problems on a separate sheet of paper.

1) What is the general equation or formula for a horizontal line given the point (a, b)?

2) Find the slope of the line and determine the equation of the line passing through $(-15, 6)$ and $(2, 6)$.

3) Graph the line $y = -30$ on a coordinate plane. Remember to scale your graph.

4) Graph the horizontal line and vertical line that goes through the point $(-4, -4)$.

5) What is the equation of the horizontal line through $(-4, -4)$?

6) What is the equation of the vertical line?

7) Find the domain and range of the line $y = 7$.

8) Write the equation of a horizontal line in function notation for function g that intersects the point $\left(3, \frac{1}{2}\right)$.

9) What is the general equation or formula for a vertical line given the point (a, b)?

10) Find the slope and the equation of the line passing through $(0, 1)$ and $(0, -5)$. Without graphing, what is the special name for this line?

11) Write the equation of the lines given in the graph. Name the intersection point.

12) What is the domain for the horizontal line?

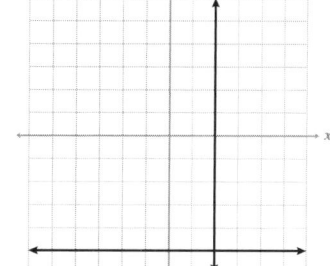

13) Write the equation of the lines given in the graph. Name the intersection point.

14) What is the domain for the vertical line?

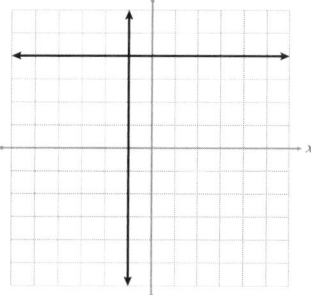

15) Determine the domain and range for the line $x = -12$.

16) Write the equation for the horizontal and vertical line that passes through the point $(-30, 15)$.

TARGETED REVIEW 11

◎ Targeted Review

In the Targeted Review, you will practice topics you have mastered in earlier lessons. Reviewing these concepts will help you be successful as you work through this unit.

Complete the problems on a separate sheet of paper.

1) Write the equation for the line (in slope-intercept form) that passes through the points $(-2, 1)$ and $(1, 5)$.

Violet was training her turtle, Mr. Fred, for a race. He was averaging 6 feet per minute and needed his current time recorded. When the timer began, Mr. Fred was 9 feet behind the starting line.

2) Write an equation in slope-intercept form to model this situation.

3) What does the rate of change mean in context?

4) If the track is 15 feet long, how long will it take Mr. Fred to finish the turtle race?

5) Graph the line, p, that travels through the point $(-3, 0)$ and has a slope of $\frac{3}{4}$.

6) Determine the equation for the line using the information in problem 5.

7) Given $b(x) = -\frac{4}{3}x$. The function $g(x)$ is translated up three units from $b(x)$. Write the function for $g(x)$ and explain why this will translate $b(x)$ up three units.

8) Write the equation for the context in function notation: Camren spent $2.50 per week on pencils for her students.

9) Explain what the domain and range are in your own words. How do they relate to the independent and dependent values?

10) Solve the compound inequality on a number line. $-19 \leq 6x - 7 < 23$

11) Set up the following conversions to find the number of pizzas which must be ordered for each student to get two slices of pizza. (Round to the nearest unit, assuming partial pizzas are not sold.)

    8 slices per pizza; 1,252 students; 2 slices of pizza per student

Multiple Choice

_____ 12) Elle borrowed $300 to start a business. She was able to earn $45 per week after expenses. How many weeks will it take for Elle to pay back her debt (owe $0)?

   A) 6 weeks

   B) $6\frac{2}{3}$ weeks

   C) 7 weeks

   D) 10 weeks

_____ 13) Determine the graph that best represents the equation: $y - 6 = \frac{2}{3}(x + 2)$

A)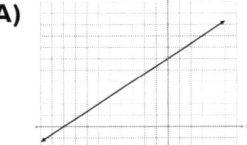

B)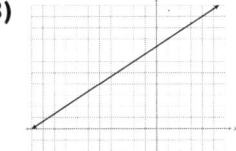

C)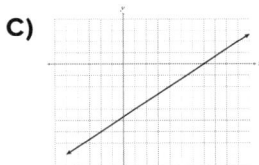

D)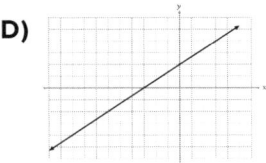

Lesson 12
Parallel and Perpendicular Lines

Outline

Part A:
Parallel Lines

- Parallel Lines
- Parallel Lines from Equations
- Parallel Lines through a Point

Part B
Perpendicular Lines

- Perpendicular Lines
- Perpendicular Lines from Equations
- Perpendicular Lines through a Point
- Special Perpendicular Lines

Targeted Review

Vocabulary

- parallel lines
- perpendicular lines

12A EXPLORE

Part A: Parallel Lines

Objectives

In this part of the lesson, you will learn about parallel lines.

By the end of this lesson, you will be able to do the following:

- ✓ Identify parallel lines.
- ✓ Write the equation of a line that is parallel to another known line and passes through a given point.

Why?

Parallel lines do not show up often in algebra. However, it is critical to understand their uniqueness when you encounter problems that include them.

☁ Warm Up

Simplify the following ratios.

1) $\frac{5}{15}$

2) $\frac{25}{60}$

3) $\frac{16}{24}$

Solve.

4) $\frac{3}{v} = \frac{9}{15}$

5) $\frac{6}{4} = \frac{r}{20}$

▶ Parallel Lines

- Parallel lines (∥) have _____ slopes, but _____ y-intercepts.

- Parallel lines have _____ points in common.

- Parallel lines can also be thought of as _____ of the same line.

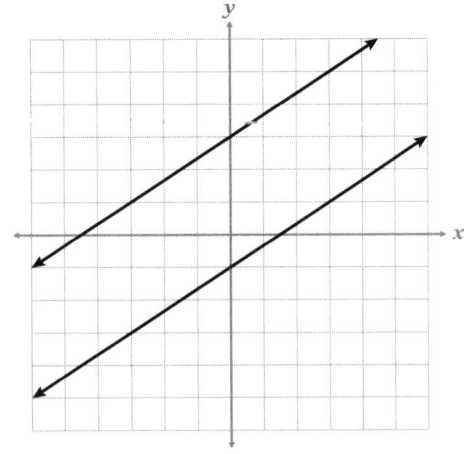

EXPLORE 12A

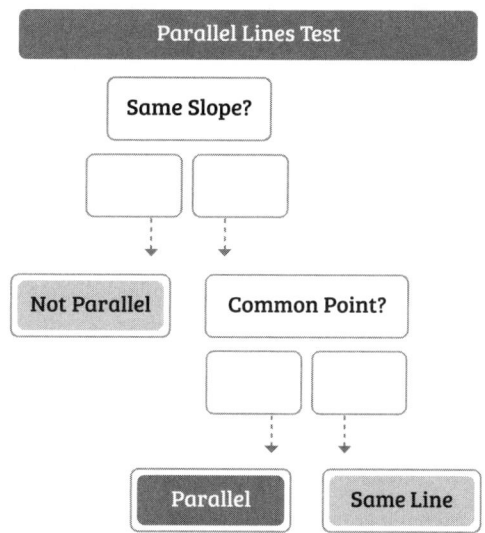

Example 1

Determine which lines represented on the graph are parallel.

Plan Determine the slope of each line. Determine the y-intercepts for lines with the same slopes.

Implement

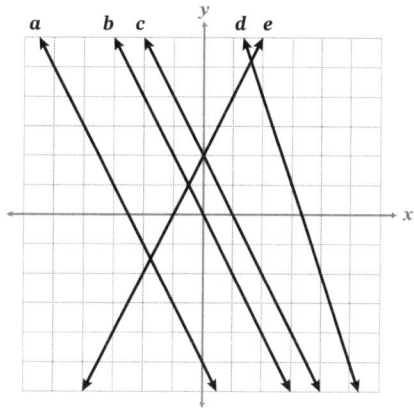

Line	m	b
Line a		
Line b		
Line c		
Line d		
Line e		

Lines a, b, and c all have the same slope and different y-intercepts. These three lines are parallel.

☑ Checkpoint

Name the slope of each line. Then determine the set(s) of parallel lines.

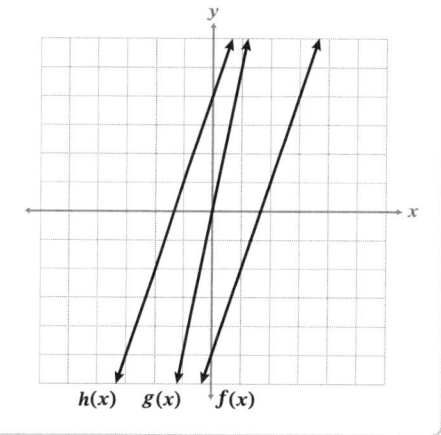

Algebra 1 Student Worktext   Lesson 12: Parallel and Perpendicular Lines > Part A: Parallel Lines > Explore   229

12A EXPLORE

▶ Parallel Lines from Equations

- _____ is particularly useful for determining if lines are parallel because you are given the slope and the y-intercept.

- Efficient ways to compare equations:
    - Write all equations in the _____.
    - Find _____ and _____.

- All horizontal lines are _____ to one another because _____ for every horizontal line and they have different y-intercepts.

- All vertical lines are parallel to one another because they have _____ slopes and different _____-intercepts.

Example 2

Determine which lines are parallel. Write your answer in a sentence, then write your answer using math shorthand.

Line m: $y = 3$

Line n: $y = 5$

Line p: $x = 3$

Line q: $x = -2$

Lines m and n are _____ lines, so _____.

Lines p and q are _____ lines, so _____.

EXPLORE 12A

Example 3

Determine which of the given lines are parallel.

Line a: $y = -\frac{2}{3}x - 1$

Line b: $2x + 3y = 6$

Line c: $y - 3 = -2(x - 2)$

Line d: $y = -\frac{2}{3}$

Line e: $x = -\frac{2}{3}$

Plan List the slopes and y-intercepts.
Use a method to compare the equations.

Implement

Line	m	b	$y = mx + b$
a: $y = -\frac{2}{3}x - 1$			
b: $2x + 3y = 6$			
c: $y - 3 = -2(x - 2)$			
d: $y = -\frac{2}{3}$			
e: $x = -\frac{2}{3}$			

Lines a and b have the same slope and different y-intercepts, so _____.

☑ Checkpoint

Are the given lines parallel? Explain.

Line a: $2x - 3y = -15$ Line b: $y = \frac{2}{3}x + 4.5$ Line c: $y = \frac{3}{2}$

Algebra 1 Student Worktext   Lesson 12: Parallel and Perpendicular Lines > Part A: Parallel Lines > Explore

12A EXPLORE

▶ Parallel Lines through a Point

- Sometimes lines need to be _____ and one of the lines must travel _____ a specific point.

- In this course, if the _____ for the solution of a linear equation is not specified, you should give the final solution in the _____ in which it was originally presented.

Example 4

Find the equation of the line in slope-intercept form that travels through the point (1, 4) and is parallel to the given line on the coordinate plane.

Plan   Find the slope of the provided line.
Write down the point for the new line.
Write the new line in point-slope form.
Write the equation of the line in slope-intercept form.

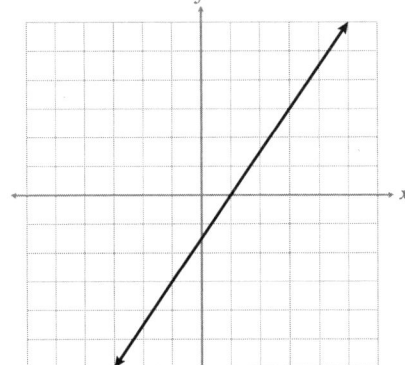

Implement

Slope of provided line: $m = \frac{3}{2}$

Point on the new line: (1, 4)

$m = \frac{3}{2} \parallel m = \frac{3}{2}$
$\phantom{m = \frac{3}{2} \parallel\ } (1, 4)$

EXPLORE 12A

Example 5

Find the equation of the line parallel to $x - 4y = -3$ that travels through the point $(-2, 7)$.

Plan   Determine the slope from the given equation.
Substitute the new point and the slope into point-slope form.
Rewrite in standard form.

Implement

☑ Checkpoint

Determine the line parallel to $y = 2x - 4$ and traveling through the point $(3, -2)$.

Practice 1

Complete the practice problems on a separate sheet of paper.

1) How do you determine if two lines are parallel?

Determine if the lines are parallel, the same, or neither.

2) $y = 2x + 1$
   $y = 2x + 4$

3) $y - 3 = 3(x + 2)$
   $y + 3 = 3(x + 4)$

4) $4x + 3y = 5$
   $4x + 3y = 7$

5)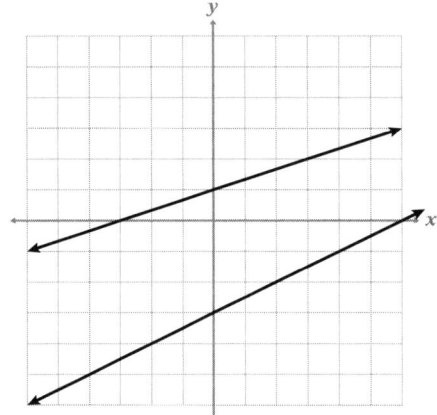

6) Given the graph of $a(x)$, $b(x)$, and $c(x)$, which, if any, of these lines are parallel? Explain.

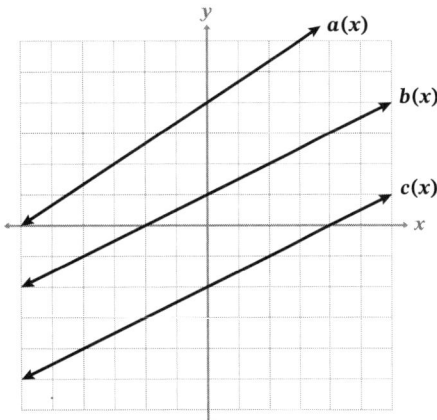

7) Determine which lines below are parallel to $2x - y = 5$.

   Line a: $4x - 2y = 7$

   Line b: $6x + 3y = 15$

   Line c: $8x - 4y = 9$

PRACTICE 1 **12A**

8) Determine if the following lines are parallel to $y - 3 = \frac{2}{3}(x + 3)$.

   Line a: $y - 5 = \frac{2}{3}(x - 3)$

   Line b: $y + 2 = \frac{2}{3}(x - 3)$

   Line c: $y - 1 = \frac{2}{3}(x - 6)$

9) Determine if the following lines are parallel to $y = \frac{1}{4}x$.

   Line a: $y - 2 = 4(x + 1)$

   Line b: $2x - 8y = 1$

   Line c: $x - 4y = 3$

Find the equation of the line that is parallel to the given line through the given point.

10) $y - 3 = 5(x + 2)$ through the point $(-4, 6)$ in point-slope form.

11) $y = -\frac{2}{5}x - 1$ through the point $(1, 1)$ in slope-intercept form.

12) $8x - 3y = 24$ through $(1, 1)$ in standard form.

13) $x = 4$ through the point $(-2, 5)$.

14) A city planner is determining the needed roads in a new neighborhood. He decides that all streets named after trees, Oak and Elm, will run parallel to one another. The center of the neighborhood is at $(5, -5)$ and Oak Street runs along the line $y = 4x + 1$. Find the equation for Elm Street that is parallel to Oak Street and goes through $(5, -5)$.

15) While laying tile for swim lanes at the bottom of a pool, the contractor noticed that the edge of the pool fit the equation of $y = -2$. The pool required four parallel lanes, each translated over four positive units. What are the equations for the next three lanes?

12A MASTERY CHECK

Mastery Check

Show What You Know

A trapezoid is a quadrilateral with one set of parallel sides.

A) State the parallel sides.

B) Prove algebraically that the lines in part A are parallel.

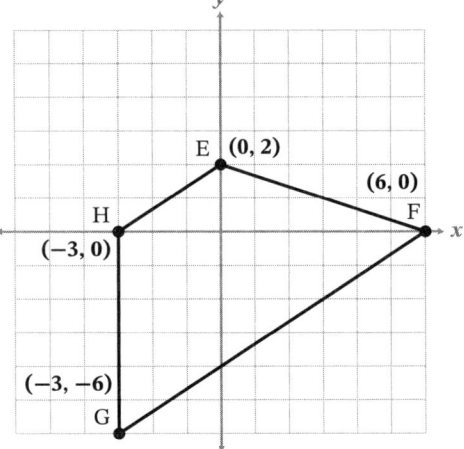

C) Find the equation of side $\overline{GH}$. Explain your thinking.

D) Connect points E and G with a segment. What is the equation of $\overline{EG}$ written in standard form?

Say What You Know

In your own words, talk about what you have learned using the objectives for this part of the lesson and your work on this page.

Practice 2

Complete the practice problems on a separate sheet of paper.

1) Given the graphs of $f(x)$, $g(x)$, and $h(x)$, which, if any, are parallel? Explain.

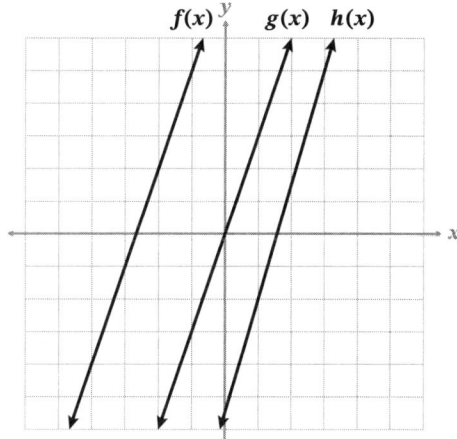

2) Determine if the following lines are parallel to the line in the graph.

Line a: $3x - y = 1$

Line b: $y + 1 = \frac{1}{3}(x - 9)$

Line c: $y - 1 = \frac{1}{3}(x - 6)$

Line d: $x - 3y = 2$

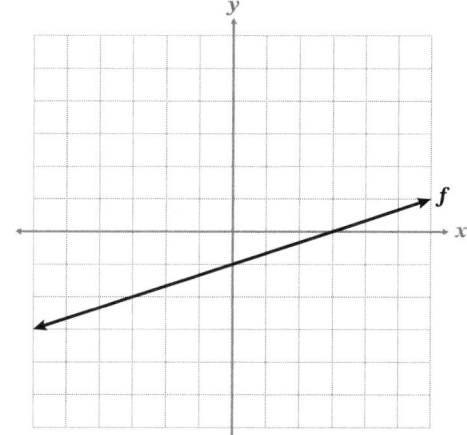

3) Determine if the following lines are parallel to $6x + 7y = 3$.

Line a: $12x + 14y = 9$

Line b: $7x + 6y = 3$

Line c: $6x - 7y = 4$

12A PRACTICE 2

4) Determine if the following lines are parallel to $y = 6$.

Line a: $y - 5 = 0$

Line b: $y = -4x$

Line c: $x = 2$

5) Determine if the following lines are parallel to $y = \frac{3}{4}x - 7$.

Line a: $y + 5 = -\frac{3}{4}(x + 3)$

Line b: $y + 5 = \frac{3}{4}(x + 3)$

Line c: $6x - 8y = 9$

6) Determine if the following lines are parallel to $y + 1 = -\frac{5}{3}(x + 3)$.

Line a: $5x - 3y = 9$

Line b: $y = -\frac{5}{3}x - 5$

Line c: $y + 1 = -\frac{5}{3}(x + 3)$

Find the equation of the line that is parallel to the given line through the given point.

7) $y + 10 = \frac{2}{3}(x + 7)$ through the point $(5, 0)$ in point-slope form.

8) $y = \frac{3}{4}x$ through the point $(8, 3)$ in slope-intercept form.

9) $4x - 3y = 6$ through $(-12, 12)$ in standard form.

10) $x = 7$ through $(-3, 1)$.

11) A gardener is making parallel rows for planting. The first row fits the equation $y = 3x - 4$. The second row must go through the point $(0, 1)$ and the third row must go through the point $(3, -2)$. Find the equations of the second and third rows in slope-intercept form.

12) A local tech company is installing a new fiber optic cable. The equation for the old line is $y = -\frac{1}{2}x - 5$. The new cable needs to run parallel to the old line but translated up three units. What is the equation for the new fiber optic cable?

Part B: Perpendicular Lines

Objectives

In this part of the lesson, you will learn about perpendicular lines.

By the end of this lesson, you will be able to do the following:

- ✓ Identify perpendicular lines.
- ✓ Write the equation of a line that is perpendicular to a given line and passes through a given point.

Why?

Perpendicular lines do not show up too often in algebra. However, it is critical to understand their uniqueness when you encounter problems that include them.

Warm Up

Write the value that is the reciprocal of the value shown.

1) $\frac{3}{2}$ reciprocal: _____
2) $\frac{1}{2}$ reciprocal: _____
3) -3 reciprocal: _____

Determine the opposite of the value shown.

4) $\frac{2}{3}$ opposite: _____
5) -2 opposite: _____

▷ Perpendicular Lines

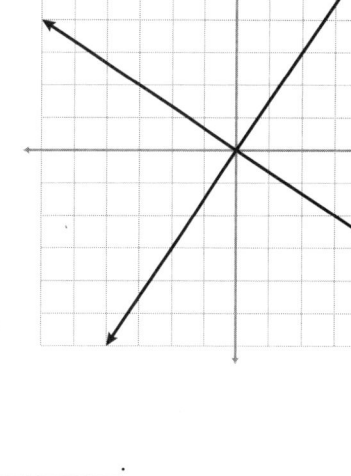

- _____ intersect each other in the same plane and form right angles.

- Slopes for perpendicular lines:
    - Have _____ signs.
    - Are _____ of one another.

- The product of the slopes of perpendicular lines will always be _____.

- Perpendicular lines can have the same _____.

- Only the _____ needs to be considered when determining if lines are perpendicular.

12B EXPLORE

Example 1

Determine the lines that are perpendicular to each other.

Slope of line b:

Slope of line c:

Slope of line d:

Slope of line a:

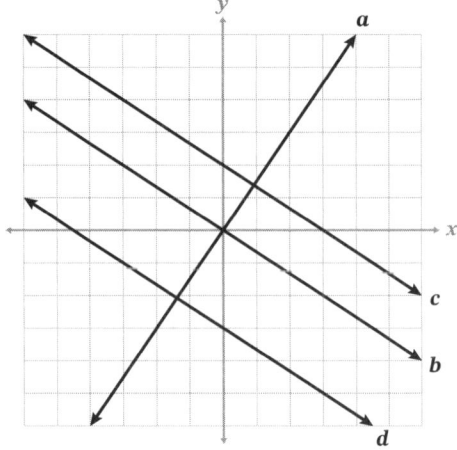

☑ Checkpoint

Determine the lines perpendicular to one another. Explain.

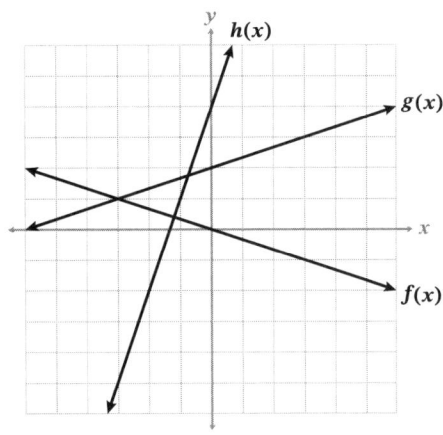

▶ Perpendicular Lines from an Equation

- The slope from _____ form or _____ form can be used to determine whether lines are perpendicular.

- If lines are given in different forms, first _____ of each line to compare.

EXPLORE 12B

Example 2

Determine which lines are perpendicular.

Line a: $y - 3 = -\frac{1}{2}(x - 2)$ Line b: $y = 3x + 4$ Line c: $x + 3y = 12$

Plan   Identify slopes.
Compare.

Implement

$m_a =$ $m_b =$ $m_c =$

> Another way to note the line you are referring to without writing out "line a" is to use m_a, which means the slope of line a.

Example 3

Determine which lines are perpendicular.

Line a: $5y + 3x = 1$

Line b: $5x - 3y = -3$

Line c: $5x + 3y = 4$

Use the formula for slope from standard form, $m = -\left(\frac{A}{B}\right)$.

☑ Checkpoint

Name the slope of each line. Determine which lines are perpendicular.

Line a: $y = \frac{1}{4}x + 2$

Line b: $y - 3 = 4(x - 2)$

Line c: $4x + y = \frac{1}{3}$

12B EXPLORE

▶ Perpendicular Lines through a Point

- Perpendicular lines can be determined given an _____ and a _____.

- First, determine the _____ of the given line.

- Then find the _____ of that slope.

- Use the new slope and the point to find a _____.

- Organize the information this way:

Example 4

Determine the line that is perpendicular to $y = \frac{3}{5}x + 1$ and travels through the point $(-1, 2)$.

Plan: Determine the slope of the new line.
Create the equation for the new line.

Implement:

☑ Checkpoint

Determine the line that is perpendicular to $x + 2y = 3$ and travels through the point $(4, 5)$. Write the equation in point-slope form.

EXPLORE 12B

▶ Special Perpendicular Lines

- The x-axis and y-axis are a set of perpendicular lines that form the _____.

- The slope of every _____ line is $m = \frac{0}{1} = 0$.

- The slope of every _____ line is $m = \frac{1}{0} =$ undefined.

- *Every* horizontal line is _____ to *every* vertical line on the coordinate plane.

Example 5

Write *horizontal* or *vertical* next to each equation. Then determine which lines are perpendicular.

Line *a*: $x = 5$

Line *b*: $y = 5$

Line *c*: $x = -2$

Example 6

Determine the line that is perpendicular to $y = -1$ and travels through the point $(-3, 5)$.

The line given is _____.

Any line _____ to this line will be _____.

The vertical line that travels through the point $(-3, 5)$ is _____.

☑ Checkpoint

Determine the line that is perpendicular to $x = \frac{2}{3}$ and travels through the point $(1, 2)$.

12B PRACTICE 1

✏️ Practice 1

Complete the practice problems on a separate sheet of paper.

Match each of the given slopes to its corresponding perpendicular slope.

Given

_____ 1) $m = 2$
_____ 2) $m = \frac{6}{5}$
_____ 3) $m = -1$
_____ 4) $m = \frac{1}{2}$
_____ 5) $m = -\frac{5}{6}$

Perpendicular Slopes

A) -2
B) 1
C) $-\frac{1}{2}$
D) $\frac{6}{5}$
E) $-\frac{5}{6}$

Determine which lines are perpendicular and which are parallel.

6)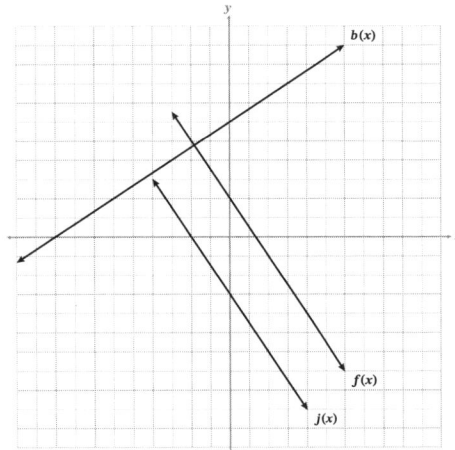

7)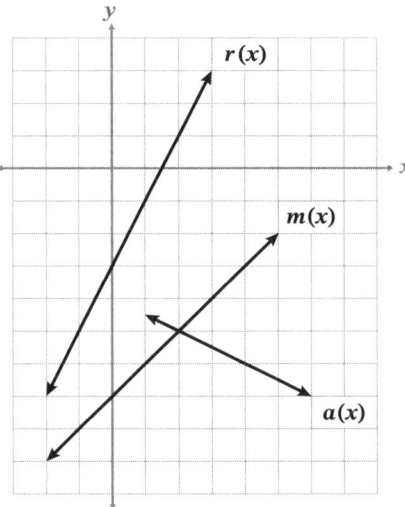

Determine which lines are perpendicular (⊥). Name the slope of each line. Write a sentence describing the relationship between perpendicular lines. Indicate any lines that are vertical or horizontal. (Hint: there can be multiple sets of ⊥ lines.)

8) Line a: $y = -\frac{3}{4}x - 5$

   Line b: $y = \frac{4}{3}x + 1$

   Line c: $y = -\frac{5}{4}x - 9$

   Line d: $y = \frac{4}{5}x + 10$

9) Line a: $4x + 3y = 12$

   Line b: $16x + 10y = 17$

   Line c: $6x - 8y = 5$

   Line d: $5x - 8y = 2$

Determine which lines are perpendicular (⊥). Name the slope of each line. Write a sentence describing the relationship between perpendicular lines. Indicate any lines that are vertical or horizontal. (Hint: there can be multiple sets of ⊥ lines.)

10) Line a: $y - 1 = -4(x + 2)$

   Line b: $y + 9 = 3(x - 8)$

   Line c: $y + 9 = -\frac{1}{3}(x - 8)$

   Line d: $y - 1 = \frac{1}{4}(x + 2)$

11) Line a: $14x - 16y = 3$

   Line b: $y = 17$

   Line c: $y - 3 = 2(x - 6)$

   Line d: $y = -\frac{8}{7}x + 5$

   Line e: $2x + 4y = 9$

   Line f: $x = 10$

Write the equation of the line that is perpendicular to the given line and passes through the given point.

12) $y = 3x - 7; (-6, 3)$ in slope-intercept form.

13) $y + 4 = \frac{1}{2}(x + 1); (-1, -4)$ in point-slope form.

14) $8x + 6y = 3; (2, -2)$ in standard form.

15) $y = 5; (-10, 11)$

16) Determine if the following lines are perpendicular to the line in the graph below. Equations a–e should be solved algebraically.

   Line a: $3x - y = 2$

   Line b: $y + 1 = -\frac{1}{3}(x - 1)$

   Line c: $3x + y = 1$

   Line d: $y + 1 = -3(x - 6)$

   Line e: $x - 3y = -3$

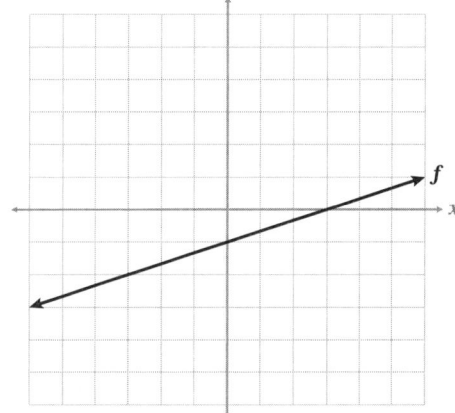

17) Renee Street and Descartes Street are perpendicular. The equation for Renee Street is $y = 5x + 1$. Descartes Street goes through the point $(15, -5)$. Find the equation of Descartes Street in slope-intercept form.

Mastery Check

✎ Show What You Know

A rectangle is a quadrilateral with special properties. Opposite sides must be parallel and adjacent sides must be perpendicular.

Prove algebraically that rectangle ABCD is, in fact, a rectangle.

A) Determine the slopes of all sides of the rectangle algebraically.

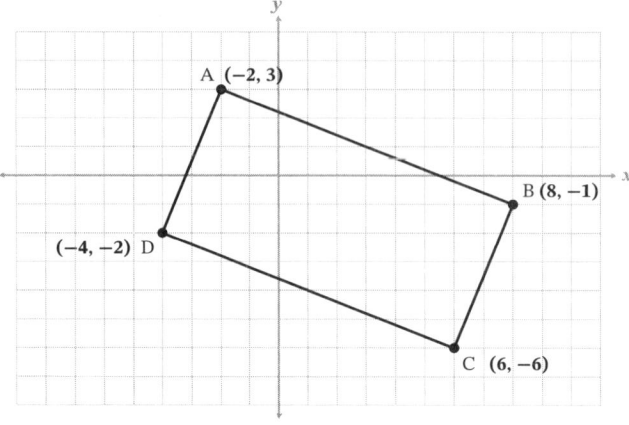

B) Name the sides parallel to one another.

C) Name the sides perpendicular to one another.

D) A segment was drawn connecting A and C to make two triangles. If the length of side $\overline{AD}$ is 5.4 cm and the length of side $\overline{CD}$ is 10.8 cm, what would the area be for one of the triangles formed by segment $\overline{AC}$?

·ıl|ı· Say What You Know

In your own words, talk about what you have learned using the objectives for this part of the lesson and your work on this page.

Practice 2

Complete the practice problems on a separate sheet of paper.

Match each of the given slopes to its corresponding perpendicular slope.

Given

_____ 1) $m = 4$

_____ 2) $m = -\frac{7}{9}$

_____ 3) $m = 0$

_____ 4) $m = -\frac{2}{3}$

_____ 5) $m = \frac{4}{7}$

Perpendicular Slopes

A) undefined

B) $\frac{9}{7}$

C) $-\frac{7}{4}$

D) $-\frac{1}{4}$

E) $\frac{3}{2}$

Find *all* pairs of perpendicular lines.

6)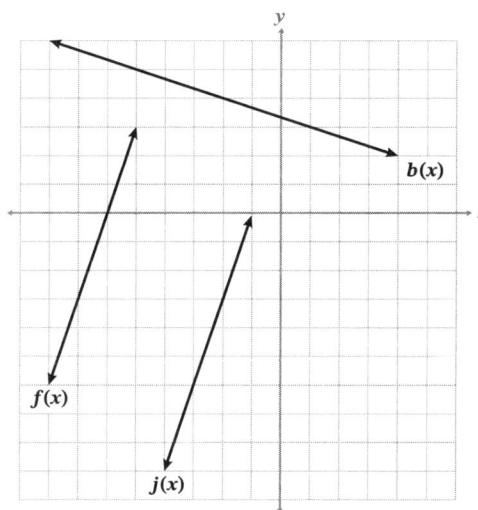

7)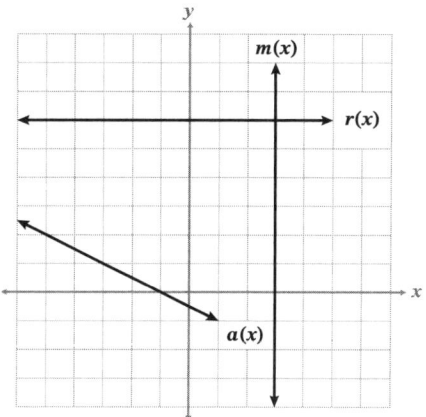

Determine which lines are perpendicular. Name the slope of each line. Write a sentence describing the relationship between perpendicular lines. Indicate any lines that are vertical or horizontal. (Hint: there can be multiple sets of ⊥ lines.)

8) Determine which lines are perpendicular.

   Line a: $y = -10x$

   Line b: $y = x - 7$

   Line c: $y = \frac{1}{10}x - 9$

   Line d: $y = -x + 10$

9) Determine which lines are perpendicular.

   Line a: $2x + 7y = 5$

   Line b: $14x - 4y = 9$

   Line c: $x + y = 5$

   Line d: $x - y = 3$

12B PRACTICE 2

Determine which lines are perpendicular. Name the slope of each line. Write a sentence describing the relationship between perpendicular lines. Indicate any lines that are vertical or horizontal. (Hint: there can be multiple sets of ⊥ lines.)

10) Determine which lines are perpendicular.

   Line a: $y - 6 = -\frac{2}{3}(x + 4)$

   Line b: $y - 1 = 4(x - 8)$

   Line c: $y - 6 = -\frac{1}{4}(x + 4)$

   Line d: $y - 1 = \frac{3}{2}(x - 8)$

11) Determine which lines are perpendicular.

   Line a: $x = 0$

   Line b: $y = -\frac{3}{5}x$

   Line c: $y = 0$

   Line d: $3x + 5y = 15$

   Line e: $y + 7 = \frac{2}{7}(x - 2)$

   Line f: $y = -\frac{7}{2}x + 9$

Write the equation of the line that is perpendicular to the given line and passes through the given point.

12) $y = \frac{5}{6}x - 7$; $(-1, 3)$ in slope-intercept form.

13) $y + 2 = 4(x + 2)$; $(8, -2)$ in point-slope form.

14) $x + 3y = 4$; $(-6, 7)$ in standard form.

15) Find the line perpendicular to the given line, $x = -12$, that passes through $(-4, 9)$. Explain your thinking.

16) Find the error, then correct the work.

   A student was trying to find the equation of a line in slope-intercept form that is perpendicular to $y = -5x + 2$ and passes through the point $(-3, 2)$. Their work was as follows:

   $y - 2 = -\frac{1}{5}(x + 3)$, or $y = -\frac{1}{5}x + \frac{7}{10}$

17) A construction worker was installing a railing for the stairs. The slope of the railing was a 7-inch rise by a 10-inch run. The worker needed to build a perpendicular baluster through the point $(-4, -6)$. Write the equation of the line in point-slope form.

Targeted Review

In the Targeted Review, you will practice topics you have mastered in earlier lessons. Reviewing these concepts will help you be successful as you work through this unit.

Complete the problems on a separate sheet of paper.

1) Graph the horizontal and vertical lines that pass through the point $(-3, 2)$. Label the point.

2) Write the equation of the horizontal line that passes through the point $(-3, 2)$.

3) Write the equation of the vertical line that passes through the point $(-3, 2)$.

4) Write an equation in point-slope form for a line with a slope of 2 that travels through the point $(-1, 3)$.

5) Graph your equation from problem 4.

6) Joseph sold lemonade for one week in the summer. He sold the lemonade for $0.75 per cup and had earned $15.25 after selling 47 cups of lemonade.

   Define your variables as an ordered pair, and write a linear equation.

   Explain what you think the y-intercept means in the context of this problem.

7) Write the equation in standard form.

   $y = -\frac{2}{3}x + 2$

8) Determine the x and y-intercepts for the given equation.

   $7x + 3y = 18$

9) Which line does each solution below satisfy?

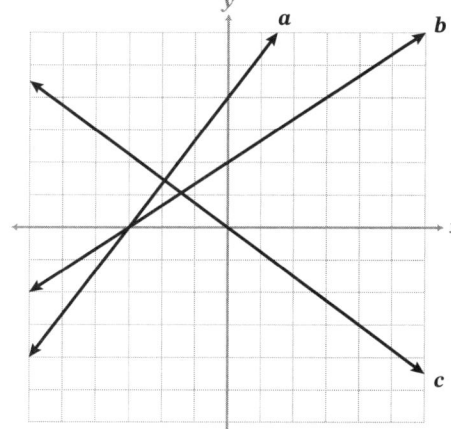

$f(0) = 0$

$f(-3) = 0$

$f(0) = 4$

$f(3) = 4$

TARGETED REVIEW 12

10) Explain how the graphs of $f(x) = -2x$ and $g(x) = -2x + 2$ are translations of one another.

11) Name the domain and range. Determine if the given relation is a function.

x	$f(x)$
-3	-9
0	0
1	3
2	6

Multiple Choice

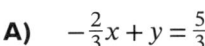 **12)** Determine the equation in standard form that represents the function $f(x)$ in the graph below.

A) $-\frac{2}{3}x + y = \frac{5}{3}$

B) $-2x + 3y = -5$

C) $2x - 3y = -5$

D) $y = \frac{2}{3}x + \frac{5}{3}$

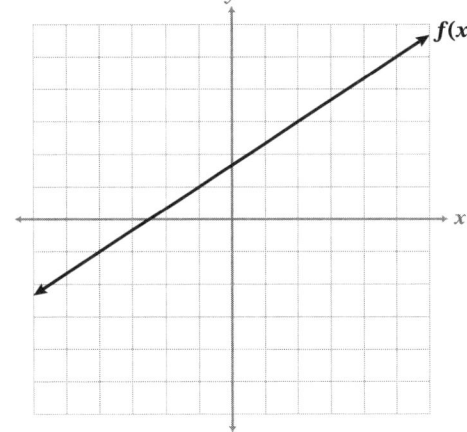

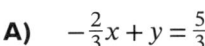 **13)** What is the equation for the horizontal line through the point $(-8, 14)$?

A) $x = -8$

B) $x = 14$

C) $y = -8$

D) $y = 14$

Lesson 13
Scatter Plots

Outline

**Part A
Correlation and Scatter Plots**

- Linear Correlations
- Graphing Scatter Plots

**Part B
Line of Best Fit**

- Calculating Line of Best Fit
- Drawing Conclusions Using the Trend Line

Targeted Review

Vocabulary

- bivariate data
- correlation
- scatter plot
- line of best fit
- interpolate
- extrapolate

13A EXPLORE

Part A: Correlation and Scatter Plots

Objectives

In this part of the lesson, you will learn about correlation and scatter plots.

By the end of this lesson, you will be able to do the following:

- Identify a correlation of a scatter plot as strong or weak, positive or negative, or no correlation.
- Explain the meanings of correlations in real-life examples.
- Create a scatter plot with accurate scale, labels, and ordered pairs.

Why?

Have you ever heard the phrase "A picture is worth a thousand words"? Scatter plots are a way to see real-life data as a picture that you can use to draw conclusions, make predictions, and prepare for likely outcomes.

Warm Up

Use the graph to classify each line by its slope (positive, negative, zero, or undefined).

1) $f(x)$

2) $g(x)$

3) $h(x)$

4) $j(x)$

5) $k(x)$

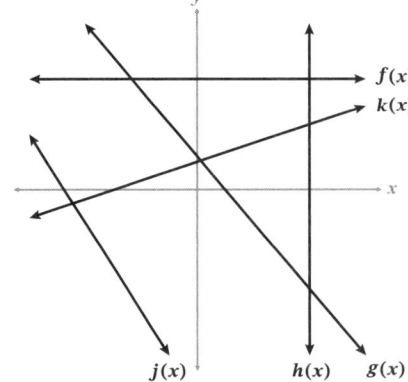

Linear Correlations

- _____ is a set of data that pairs independent and dependent variables to form a correlation.

- _____ is a relationship between two variables or two data sets.

- A _____ is a graph that visually represents bivariate data.

- A _____ (r) is the number that shows how correlated, or related, the data set is.

    - The r-value will be given to you, or will be calculated by using technology.

Positive Correlation

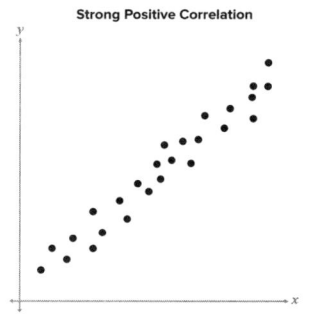

$r = 0.96$

As x _____,

y _____.

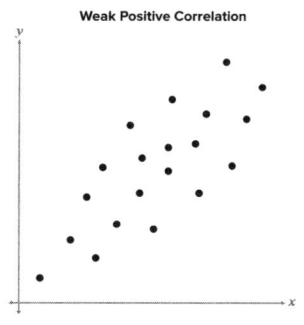

$r = 0.42$

As x increases, y increases, but the ordered pairs are more spread apart.

Negative Correlation

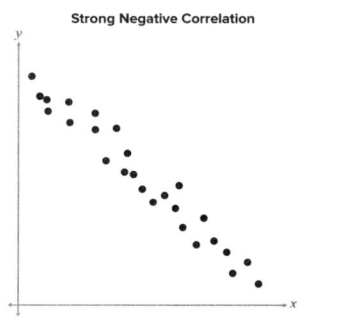

$r = -0.9$

As x _____,

y _____.

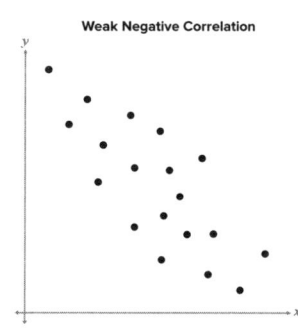

$r = -0.42$

As x increases, y decreases, but the ordered pairs are more spread apart.

No Correlation

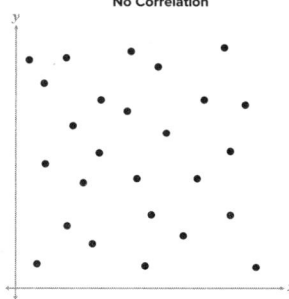

$r = 0$

The r-value equals zero.

As x _____, y has

_____ to x.

To describe correlation from a _____:
- List independent and dependent variables as _____.
- Complete sentence: As x increases, y [increases / decreases / has no relation to x]

13A EXPLORE

Example 1

Determine the correlation for the graph. If it has a positive or negative correlation, describe the meaning of the correlation in context.

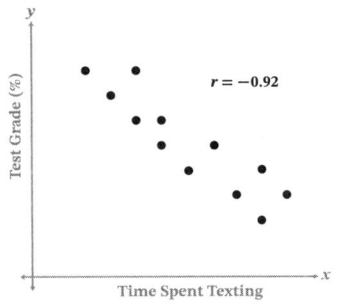

Correlation coefficient:

Independent and dependent variables:
(time spent texting, test grade)

As time spent texting _____ the test grade percentage _____ .

If a student wants to improve their test score, they could try spending less time texting.

Example 2

Name and explain the correlation for the graph.

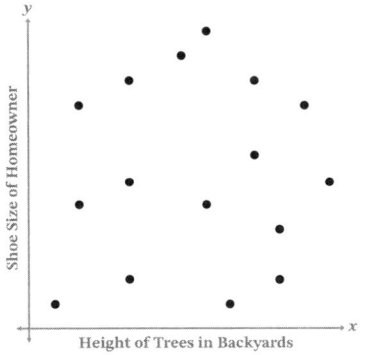

correlation coefficient:

(_____ , _____)

Because the height of trees and the shoe size of the homeowner are not related to each other,

_____ exists.

☑ Checkpoint

Determine the correlation for the graph. If it has a positive or negative correlation, describe what the correlation means within the given context.

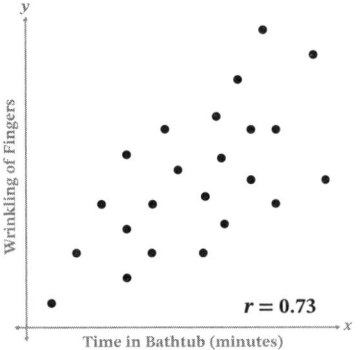

Graphing Scatter Plots

- When graphing scatter plots, it is important to graph _____ precisely.

- Before graphing the ordered pairs, determine the scale of the graph and _____ the x and y-axis.

- The ordered pairs for a scatter plot will be given as a table or a _____.

Example 3

Graph the given data set to determine the relationship between the two values.

Molly collected data from children in her neighborhood. She measured their height and their arm span in centimeters and recorded the data in a table. She wants to know the relationship, if any, between the two values. She decided to make a scatter plot.

Height (cm)	Arm Span (cm)
102	101
104	106
111	108
98	100
108	110
118	115
116	116
110	113
112	112
110	110
105	104

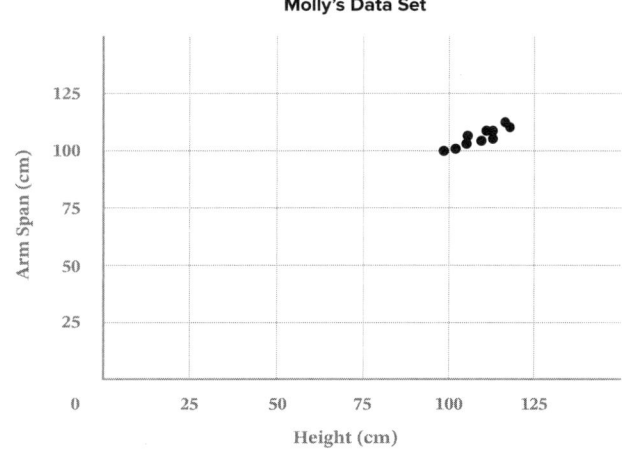

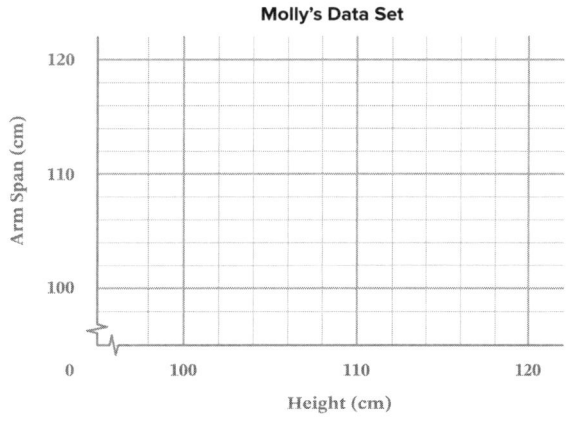

- You can use a squiggle on one or both axes to denote a _____ at the start of a graph.

- As height _____, arm span also _____.

- This looks like a _____ correlation between height and arm span.

13A EXPLORE

☑ Checkpoint

Correctly label the scatter plot describing the relationship between hair length in inches and the amount of shampoo used in teaspoons. Name and explain the correlation.

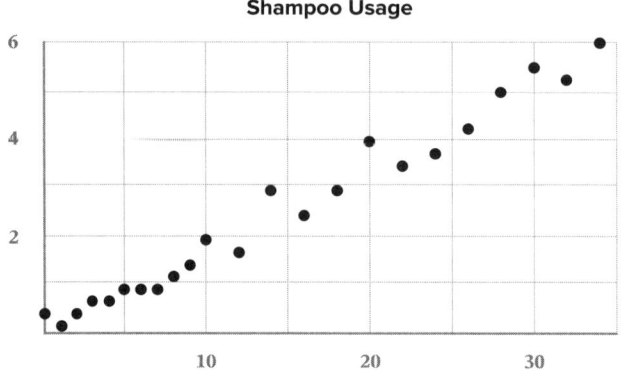

Practice 1

Complete the problems on a separate sheet of paper.

Given the type of correlation, complete the sentence and sketch an example.

1) Strong negative correlation, $r = -0.89$ As x _____, y _____.

2) No correlation, $r = 0$ As x _____, y _____.

3) Weak positive correlation, $r = 0.46$ As x _____, y _____.

4) Determine the correlation for the graph. If it is a positive or negative correlation, describe the meaning in context.

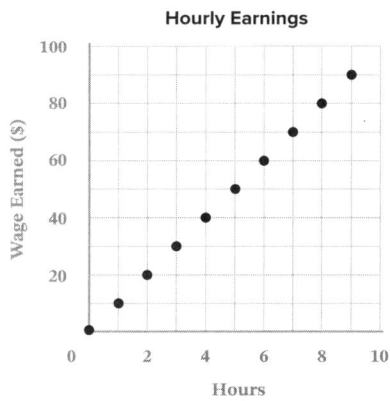

5) Determine the correlation for the graph. If it is a positive or negative correlation, describe the meaning in context.

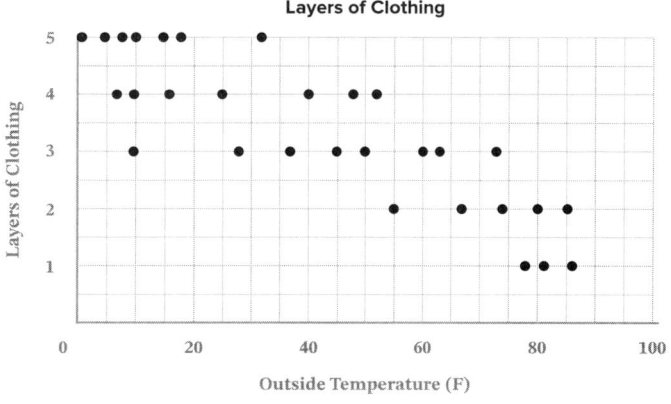

6) Emeril collected data from his class. He asked each person in his class their height and the grade they earned on their last test. Name the correlation and describe the relationship. Explain why your statement makes sense.

Use the data set to compare time spent working out in hours, x, to the resting heart rate, y, of 28-year-olds.

x	0	0.25	0.25	0.5	1	1.25	3	3.5	4.5	5.25	6	6.25	7	8
y	80	82	78	75	75	70	67	64	60	61	52	55	50	48

7) Graph.

8) Explain the scale of the graph.

9) Name the type of correlation and explain the relationship between the data sets.

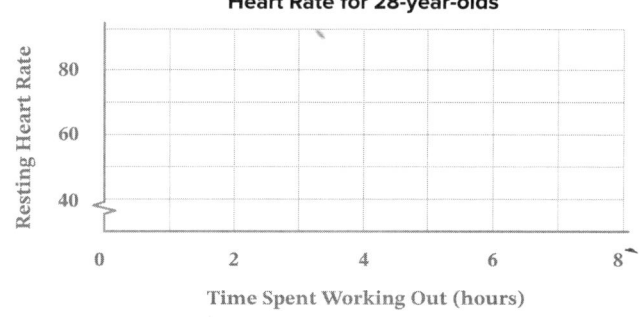

This data set compares the number of tickets sold monthly, x, to the ticket sales, y, for a local theater.

Numbers of tickets sold monthly	Ticket sales ($)
814	9,390
767	8,838
659	7,340
737	8,713
688	7,475
671	8,138
661	7,763
558	6,330
587	6,693
594	6,805
549	6,240
543	6,038

10) Graph.

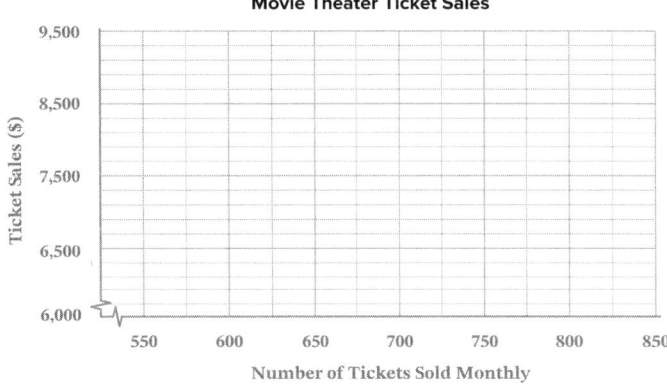

11) Explain the scale of the graph.

12) Name the type of correlation and explain the relationship between the data sets.

Mastery Check

Show What You Know

A) Create a scatter plot for the data set.

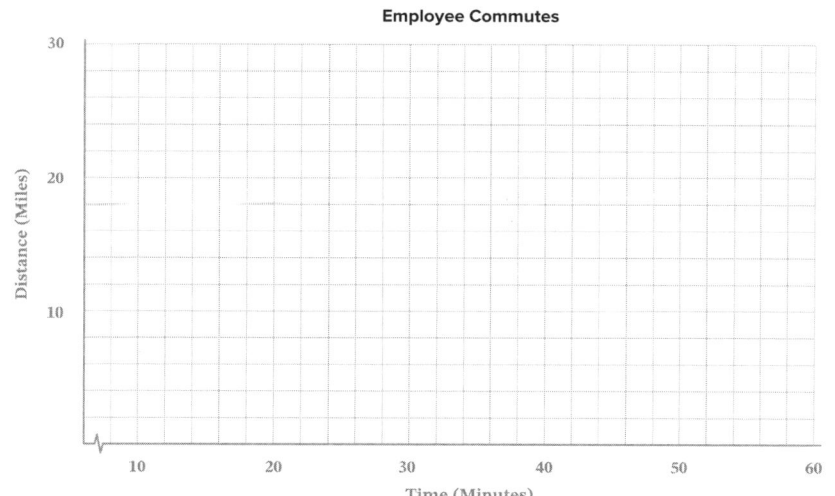

Time spent traveling to work	Distance traveled
20	15
30	21
26	26
48	25
10	7
8	6
11	7
15	10
25	17
10	10
60	30
55	30
12	8

B) Name the type of correlation and explain the relationship between the values.

C) What does the ordered pair (25, 17) mean in the context of this problem? Explain your thinking.

D) One employee reported that it took them 25 minutes to travel 2 miles to work. This person walks to work while the other employees surveyed drive. Explain why this data point was not included.

Say What You Know

In your own words, talk about what you have learned using the objectives for this part of the lesson and your work on this page.

13A PRACTICE 2

✏ Practice 2

Complete the problems on a separate sheet of paper.

Choose the graph that best matches the description. Label each axis.

A) 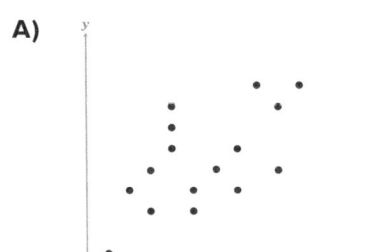   B) 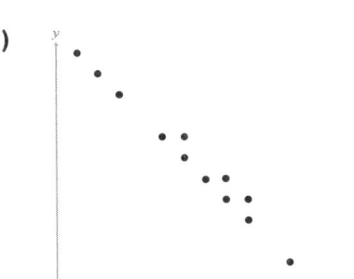   C)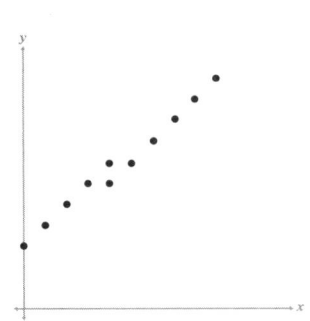

_____ 1) As the weight of the vehicle increases, the gas mileage decreases.
(weight of vehicle, gas mileage) $r = -0.97$

_____ 2) As time in weeks increased, the weight of the puppy increased.
(weeks, weight of puppy) $r = 0.98$

_____ 3) As time on a computer increased, typing speed increased.
(time on computer, typing speed) $r = 0.48$

Explain the relationship, if any, between the given variables.

4) (distance driven in miles, gas remaining in tank)

5) (height in inches, number of pairs of socks)

6) (age of a dishwasher, number of repairs needed)

Use the data set to compare the year and approximate world population in billions.

7) Graph the data set.

World Population Starting in 2000

(y-axis: Population in Billions, x-axis: Years Since 2000)

Year	Population in billions
2020	7.8
2019	7.7
2018	7.6
2017	7.5
2016	7.5
2015	7.4
2014	7.3
2013	7.2
2012	7.1
2011	7
2010	7
2009	6.9
2008	6.8
2007	6.7
2006	6.6
2005	6.5
2004	6.5
2003	6.4
2002	6.3
2001	6.2
2000	6.1

8) Describe the correlation, if any exists.

9) Would you expect the trend you described in Problem 8 to continue? Explain.

Use the data set to compare students' GPAs, x, to their SAT scores, y.

10) Graph the data set.

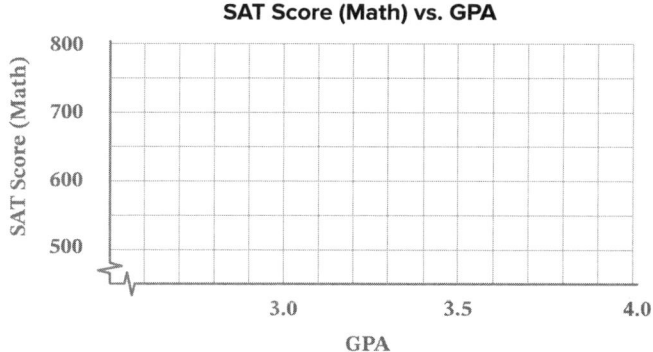

Student	GPA	SAT Score (Math)
1	3.3	600
2	3.1	530
3	2.7	500
4	3.8	750
5	2.6	500
6	3.6	660
7	3.9	750
8	3.5	680

11) Determine the type of correlation and describe its meaning in the situation.

12) Explain how to improve the graph of the data set.

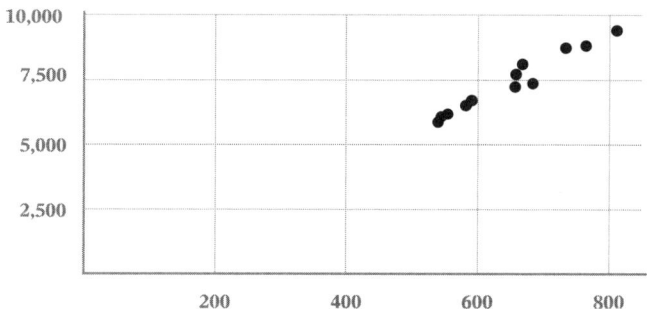

13B EXPLORE

Part B: Line of Best Fit

Objectives

In this part of the lesson, you will learn about the line of best fit.

By the end of this lesson, you will be able to do the following:

- ✓ Estimate and draw the line of best fit for a set of data.
- ✓ Write the equation for the line of best fit in the requested form.
- ✓ Use the line of best fit to interpolate, extrapolate, and explain the context of the data.

Why?

Being able to create a line of best fit for a scatter plot can help you make better use of the data. The line of best fit brings order and meaning to the scatter of ordered pairs.

Warm Up

Write the equation of the line in slope-intercept form. Round values to the nearest hundredth where necessary.

1) (2,001, 6.2), (2,011, 7)

2) (0.3, 5), (0.6, 12)

▶ Calculating the Line of Best Fit

- A _____ can be drawn on a scatter plot that approximates the bivariate data set.

- The line of best fit is a line that passes _____ the most number of data points in the set.

- When possible, there should be the _____ number of points _____ and _____ the line.

- The line of best fit is sometimes referred to as a _____ because it represents the trends, or general tendencies, between the two data sets.

Example 1

Given the scatter plot, draw the line of best fit.

The line of best fit
- passes through or close to several points
- leaves a similar number of points above and below the line

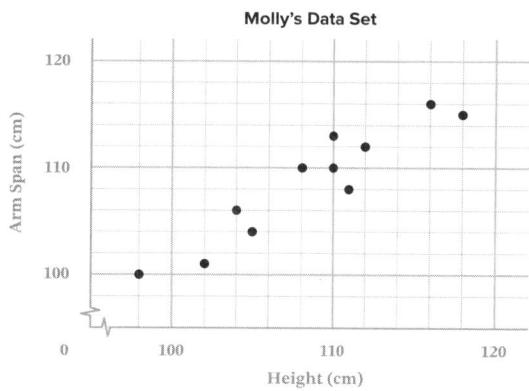

Find the line of best fit by hand:

1) Draw a scatter plot with correct labels and scale.
2) Estimate the line of best fit.
3) Pick two points on the line.
4) Find the equation of the line.

Find the line of best fit and correlation coefficient using technology:

1) Enter all ordered pairs from the data set.
2) Use the regression equation $y_1 \sim mx_1 + b$.
3) Record slope, y-intercept, and r-value.
4) If you believe outliers are present, try removing them to see if this improves the r-value.

Example 2

Molly chose two points on her line of best fit: (98, 100) and (110, 110).

Molly uses her chosen points to find the slope.

$$m = \frac{110-100}{110-98} = \frac{10}{12} = \frac{5}{6}$$

Next, Molly uses slope and the point (110, 110) to write the equation in point-slope form, then converts to slope-intercept form.

$y - y_1 = m(x - x_1)$

$y - 110 = \frac{5}{6}(x - 110)$

$y - 110 = \frac{5}{6}x - \frac{550}{6}$

$y = \frac{5}{6}x - \frac{550}{6} + 110$

$y = \frac{5}{6}x - 91\frac{2}{3} + 110$

$y = \frac{5}{6}x + 18\frac{1}{3}$

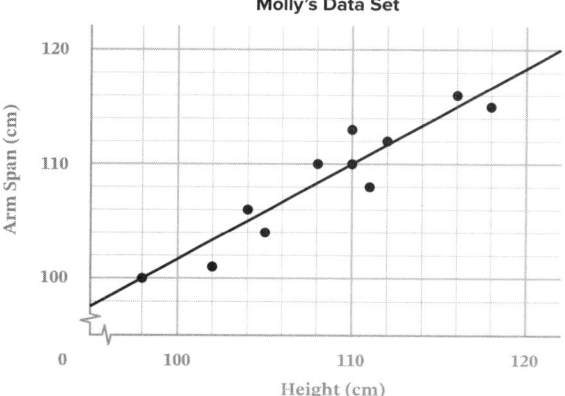

Using Technology:

If using technology, the ordered pairs and the regression equation would be entered. The values will need to be rounded to the hundredth.

$y_1 \sim mx_1 + b$

$m = 0.85, b = 16.17, r = 0.94$

Keep Molly's equation, $y = 0.85x + 16.17$, in mind because you will use it later to predict other values for the children in her neighborhood.

13B EXPLORE

Example 3

A tutor decided to compare her students' GPAs to their SAT scores. She wanted a line of best fit to generalize the trend in her data.

Plan Find the equation for the line of best fit.
- Select two points on your line, through grid lines.
- Use the slope formula to find the slope of the line.
- Use point-slope form to convert the equation into slope-intercept form.

By Hand: (675, 3.5), (550, 3.0)

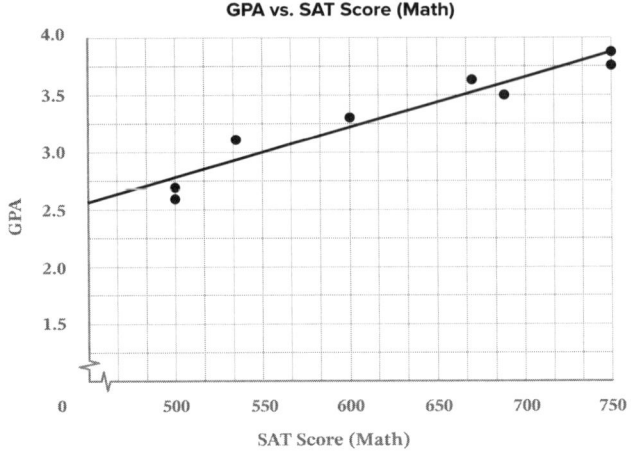

Using Technology:

$y = 0.004x + 0.537$

$r = 0.96$

☑ Checkpoint

Given the Graph: Estimate and draw the line of best fit. Then write the equation for the line in slope-intercept form using two points marked on your line of best fit.

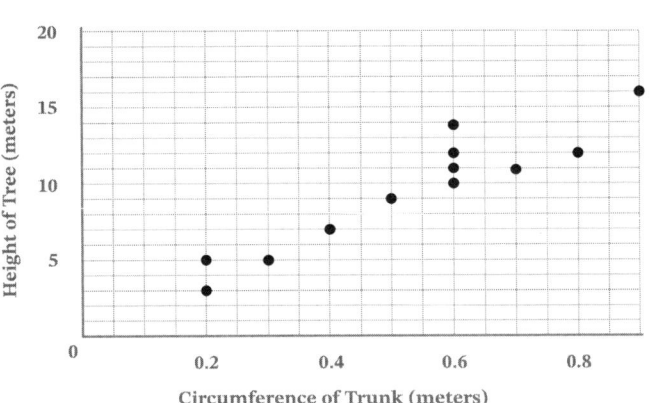

Drawing Conclusions Using the Trend Line

- Having an equation for the trend line allows you to make _____ about other values in the data set.

- _____ is the process of predicting data _____ the given points.

- _____ is the process of predicting data _____ the given points, before or after the data set.

Example 4

Use Molly's graph and equation, calculated with technology, to interpolate and find the missing data.

Molly knows the height of a child that she did not collect data from is 100 centimeters. Molly uses her equation to find the arm span of the child.

Molly's equation: $y = 0.85x + 16.17$

New data point:

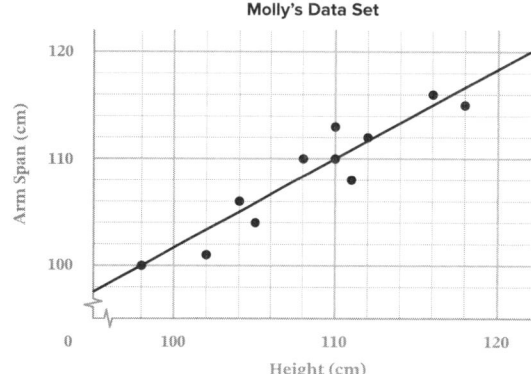

Molly's Data Set

Enter the known data into the equation:

Explain

When the child's _____ is 100 cm, the _____ is about 101 cm.

13B EXPLORE

Example 5

Molly needed to determine heights for two children. She measured their arm spans but forgot to list their heights. The arm spans were 122 centimeters and 96 centimeters. This data cannot be found on the graph, so the equation must be used to extrapolate the information.

Molly's equation: $y = 0.85x + 16.17$

A) New data point:

Enter the known data into the equation:

$122 = 0.85x + 16.17$

$105.83 = 0.85x$

$x = 124.51$

B) New data point:

Enter the known data into the equation:

Explain

☑ Checkpoint

Use the provided equation for the line of best fit to make the indicated predictions. Round to the nearest whole number.

$y = 0.5x + 4.4$ (time traveling to work, distance traveled)

A) How long will it take a person to travel 40 miles?

B) What distance was traveled when it took 45 minutes to get to work?

Practice 1

Complete the problems on a separate sheet of paper.

Write equations in slope-intercept form.

The scatter plot represents the data a movie theater collected comparing ticket sales and the number of tickets sold monthly.

1) Draw the line of best fit and mark two ordered pairs on the line.

2) Write the equation of your line using the ordered pairs.

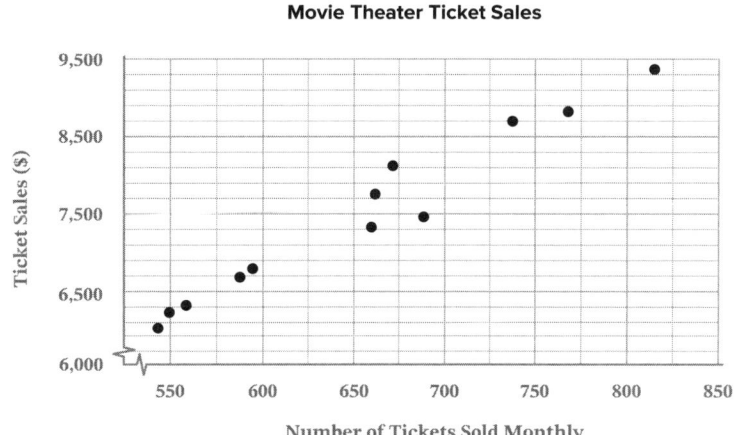

The scatter plot represents the number of teachers hired compared to the number of students in the school.

3) Draw the line of best fit and mark two ordered pairs on the line.

4) Write the equation of your line using the ordered pairs.

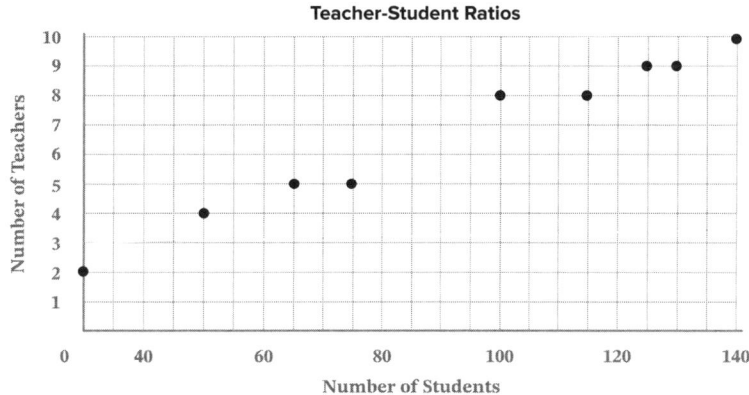

The scatter plot shows the comparison between the number of cars on the highway and the total snowfall in inches.

5) Draw the line of best fit and mark two ordered pairs on the line.

6) Write the equation of your line using the ordered pairs.

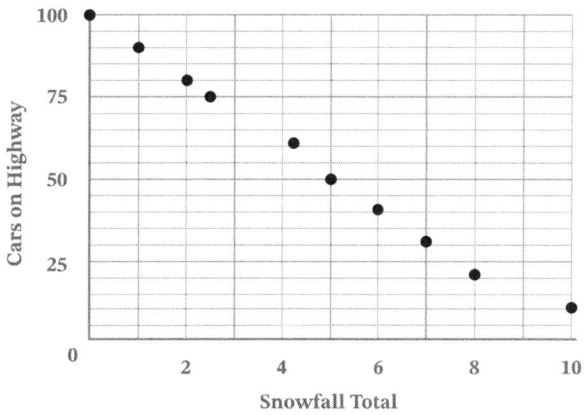

13B PRACTICE 1

Isaiah tracked his free time versus his work time on the graph below. He noticed there was a strong correlation between the two data sets.

Isaiah's Free Time Hrs vs. Work Hrs

7) Draw the line of best fit and write the equation. Show your work.

8) Using problem 7, explain the meaning of the slope and y-intercept.

Using Isaiah's equation, predict the following:

9) The number of hours of work if Isaiah has 9.5 hours of free time.

10) The number of hours of work if Isaiah has 12 hours of free time.

11) The hours of free time if Isaiah works 16 hours. What is this ordered pair called?

12) Can this equation be used to predict the number of work hours when Isaiah takes 18 hours of free time? Explain your result using numerical data.

Joel recorded his family's average monthly electric bill over the last eight years in the table below and calculated the equation $y = 2x + 130.50$.

year	2013	2014	2015	2016	2017	2018	2019	2020
bill	$131	$133	$134	$137	$138	$140	$143	$146

13) Approximately how much will Joel's family pay in 2021 for electricity? Explain if this is interpolation or extrapolation. (Hint: Find the number of years past 2013.)

14) Explain the meaning of the slope and y-intercept for Joel's equation.

Mastery Check

Show What You Know

Given the scatter plot, compare the weight of a pencil to the length of a pencil in centimeters.

A) Draw the line of best fit. Calculate the line of best fit in slope-intercept form.

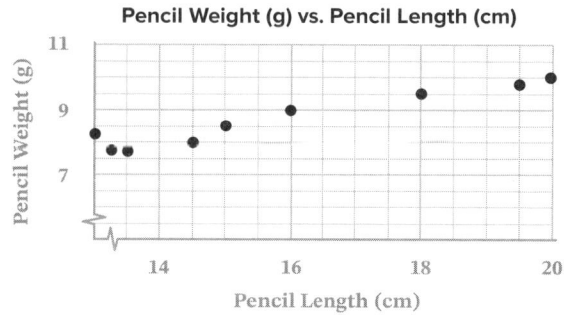

B) Explain the meaning of the slope and y-intercept.

Use your equation to make predictions about the data.

C) Predict the length of a pencil when the weight is 9 grams. Round to the nearest whole number.

D) Predict the weight of a pencil with a length of 10 centimeters. Round to the nearest whole number.

Say What You Know

In your own words, talk about what you have learned using the objectives for this part of the lesson and your work on this page.

13B PRACTICE 2

✎ Practice 2

Complete the problems on a separate sheet of paper.

1) The scatter plot below compares total steps in miles to the day of the month. Draw the line of best fit and mark two points on the line.

2) Write the equation for the line of best fit using your ordered pairs.

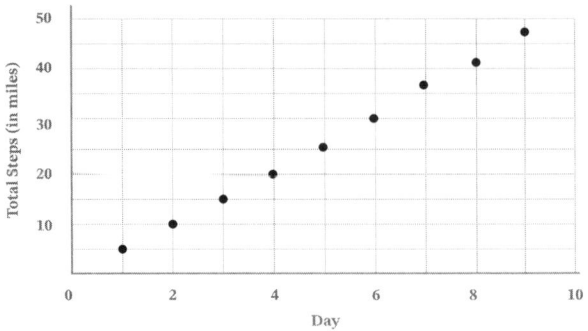

The scatter plot shows a record of Grant's reading. He compared the pages remaining in the book to the total number of pages he had read.

3) Draw the line of best fit and mark two points on the line.

4) Write the line of best fit using your points.

5) How many pages are in the novel Grant is reading? Explain.

6) Explain what the equation of the line of best fit can be used for.

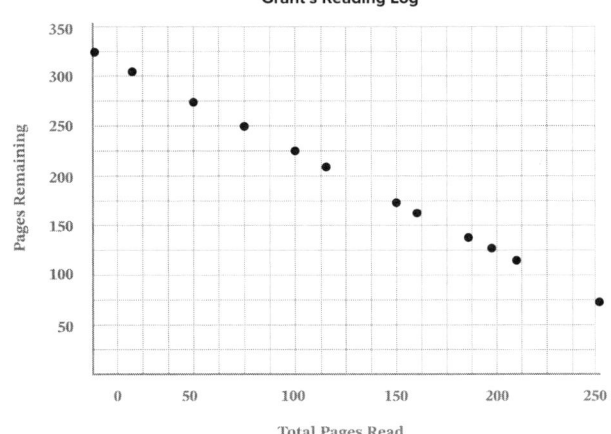

The Goldberg family went fishing once a week in the spring. They recorded the water temperature and the number of fish the entire family caught each trip. For each fishing trip, they stayed for three hours.

Water temperature (F)	Fish caught
70	5
71	8
71	5
73	6
74	5
75	5
75	4
77	3
78	2
80	1
80	0

7) Draw the scatter plot and line of best fit.

8) Write the equation for the line of best fit using the ordered pairs.

9) Explain the meaning of the slope for the Goldberg family.

Fred has a dog walking business. He charged the same amount for each dog walked and wanted an equation to calculate prices quickly.

10) Draw and write an equation to find the average charge for a dog walk.

11) What would the charge be for Fred to walk 11 dogs? Is this interpolation or extrapolation?

12) Explain the meaning of the slope and the y-intercept for Fred's dog walking business.

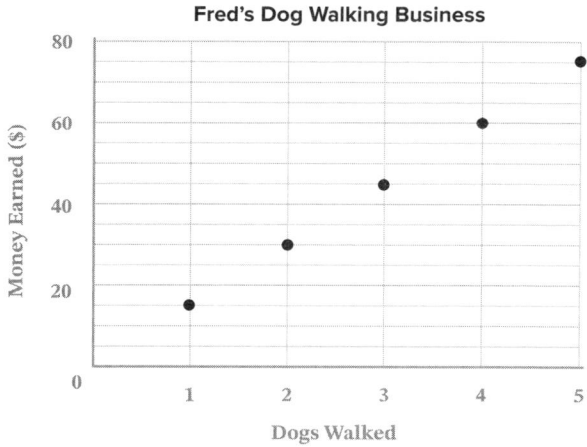

TARGETED REVIEW 13

◎ Targeted Review

In the Targeted Review, you will practice topics you have mastered in earlier lessons. Reviewing these concepts will help you be successful as you work through this unit.

Complete the problems on a separate sheet of paper.

1) Write the equation of a line parallel to $y = \frac{2}{3}x - 8$, through the point $(5, -2)$.

2) Write the equation of the line perpendicular to $f(x) = -2x$ through the origin. Name this new line $g(x)$.

3) Graph $f(x)$ and $g(x)$ from problem 2.

During the fall, an oak tree will lose around 150 leaves every 3 days. An elm tree will lose around 80 leaves every 2 days. Assume that the trees lose their leaves at a constant rate.

4) Would the slopes for these functions be positive or negative? Why?

5) Which graph would have a steeper slope?

6) The price of gasoline is $4.25 for g gallons of gas. Write a function rule for the cost of the gasoline $c(g)$. Find the price for 10.8 gallons of gas.

7) Find the equation of a line through $(-5, -5)$ and $(-1, -1)$. Write the equation in standard form.

Jessica charges a flat fee of $15 for travel and $50 an hour for a photography session.

8) Write the equation of the line in slope-intercept form.

9) Graph the line using an x scale of 1 and a y scale of 5. Will the x and y values be negative?

10) How much does Jessica charge for 2.5 hours of work?

Graph each of the linear functions on the same coordinate plane.

11) $h(x) = -\frac{3}{5}x - 2$ 12) $j(x) = 4x + 3$ 13) $v(x) = -x$

14) Use the equation $x - 2y = 4$ to find the x and y-intercepts from the equation and the slope of the equation.

15) Graph the equation $x - 2y = 4$.

Multiple Choice

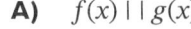 16) Determine the functions that are parallel to one another from the graph.

   A) $f(x) \| g(x)$
   B) $f(x) \| h(x)$
   C) $g(x) \| h(x)$
   D) cannot be determined

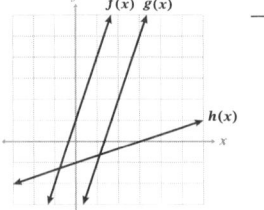

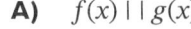 17) Name the slope perpendicular to the given equation: $3x - 6y = 7$

   A) $-\frac{1}{2}$
   B) -2
   C) $\frac{1}{2}$
   D) 2

Lesson 14
Types of Functions and Arithmetic Sequences

Outline

Part A
Continuous and Discrete Functions

- Interval Notation
- Continuous or Discrete Functions
- Functions in Various Forms

Part B
Arithmetic Sequences

- Arithmetic Sequences

Targeted Review

Vocabulary

- interval
- interval notation
- discrete function
- continuous function
- arithmetic sequence
- terms of the sequence

14A EXPLORE

Part A: Continuous and Discrete Functions

Objectives

In this part of the lesson, you will learn about continuous and discrete functions.

By the end of this lesson, you will be able to:

- Use interval notation to define the domain and range of functions.
- Decide if a function is discrete or continuous and explain the difference.
- Choose the most appropriate form of the equation of a line for a given scenario.

Why?

Will any point, integer or decimal, on a line be true? Can only whole numbers answer the question? Knowing if a situation is discrete or continuous will allow you to determine if your answers make sense for a real life scenario.

Warm Up

Use the relation to answer the following questions.
{(1, 3) (4, −2), (9, 1), (3, 4), (5, 3)}

1) Name the domain of the relation.

2) Name the range of the relation.

3) Describe the similarities and differences between the two graphs.

Graph A

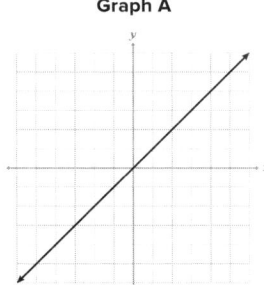

Graph B

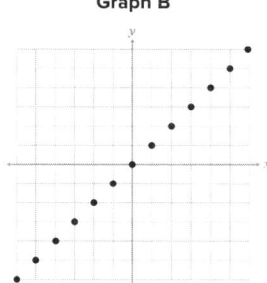

EXPLORE 14A

▶ Interval Notation

- An _____ is the set of all values between two numbers.

- _____ is one way to express every element in an interval.

- If the boundary point is _____, a closed point is used on the graph and a bracket is used in interval notation.

- If the boundary point is _____, an open point is used on the graph and a parenthesis is used in interval notation.

- This number line represented in interval notation is: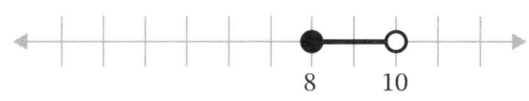

  _____

- The boundary that represents all real numbers is: _____

- Use _____ to represent the domain and range of a _____.

Example 1

For the function, $y = x^2$, write the domain and range in interval notation.

domain	range

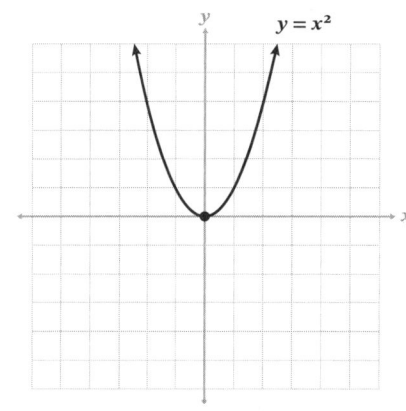

Using interval notation:

domain:

range:

This graph is in Quadrants _____.

14A EXPLORE

Example 2

For the function $y = -3x + 2$, represent the domain and range in interval notation.

x-coordinate:

y-coordinates:

domain:

range:

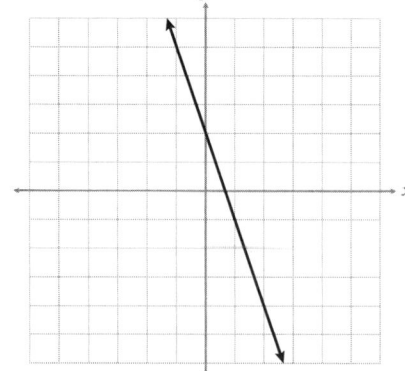

☑ Checkpoint

Provide the domain and range in interval notation for the graph below. Explain whether the graph is a function.

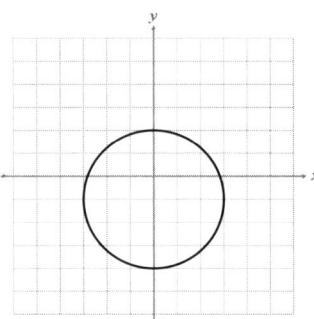

▶ Continuous or Discrete Functions

- A _____ has continuous domain elements or a domain written as a single interval.

- The graph of a continuous function will be a _____.

- _____ have a domain made up of distinct elements that can be plotted as separate points.

- The graph of a discrete function is created from a set of _____.

- This graph represents a _____ function.
    - domain: {−2, −1, 0, 1, 2}
    - range: {−5, −3, −1, 1, 3}

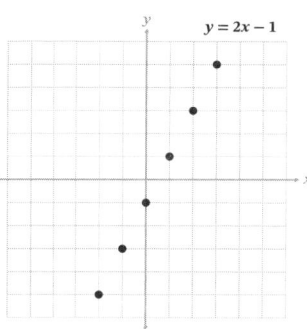

- This graph represents a _____ function.
    - domain: (−∞, ∞)
    - range: (−∞, ∞)

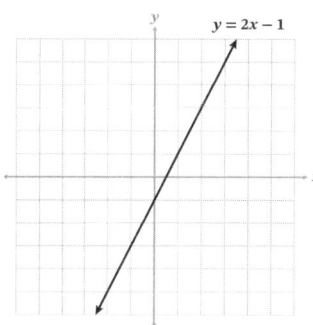

> Now that you know there are different types of functions, even within linear graphs, you will be able to more accurately portray the information.

- For both discrete and continuous _____, the independent values are associated with the _____ and the dependent values are associated with the _____.

Example 3

Randy was going to deliver a van to a family in Oklahoma. The following function shows the number of gallons, $g(m)$, for a given number of miles driven, m.

$$g(m) = \frac{m}{20}$$

A) If Randy was going to travel 1,374 miles to deliver the van, how many gallons of gasoline would he need?

(miles driven, gallons of gas used) $g(1{,}374) = \frac{1{,}374}{20}$

$m = 1{,}374$ miles

B) If he was given a gas card for 14 gallons of fuel, how many miles could he expect to drive using the gas card?

$g(m) = 14$ gallons $14 = \frac{m}{20}$

Explain

This is a _____ function, because it is possible to determine the amount of gas used for any number of miles, and gasoline is used continuously while driving.

The domain is $0 \leq m \leq 1{,}374$, or the interval _____.

The range is $0 < g(m) \leq 68.7$, or the interval _____.

14A EXPLORE

Example 4

At East-West Printers, the cost to print a company logo on an 8 × 10 inch sticker is $7.25.

A) Write the given equation, $c = 7.25p$, in function notation. Use the function to find how much it costs to print a logo a maximum of 13 times.

Cost (c) is determined by the number of logos printed (p)

(independent, dependent)   Function Notation: $c(p) = 7.25p$

To print 13 logos: $c(13) = 7.25(13)$

$= \$94.25$

B) Name the domain and range, and what each represent.

Domain: {0, 1, 2, 3, ..., 13}. The domain represents the _____.

Range: {0, 7.25, 14.50, 21.75, ..., 94.25}.

The range represents the _____.

C) Does this represent a discrete or continuous function? Explain.

The logo company would not get paid for printing part of a logo. Since the logo company only gets paid for printing a _____ of logos, the function is _____.

> In a function, ellipses are used in the domain and range to show that the pattern will continue without having to list all of the terms for the domain and range.

☑ Checkpoint

Is the graph of the following function continuous or discrete? Explain.

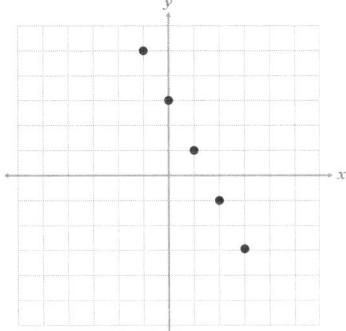

EXPLORE 14A

▶ Functions in Various Forms

- When no linear form is specified, it is up to _____ to determine the most efficient way of writing a linear equation.

- Besides choosing the best linear form to represent information, you will also need to determine if a _____ will be discrete or continuous.

Example 5

Write an equation using the most appropriate form of a line (slope-intercept, point-slope, or standard) for the given scenario. Explain why the graph would be discrete or continuous.

John was working to buy a car. He earned $12 per hour, and he already had some money that his mom had saved to help him. After working 25 hours, John had $2,800 available to purchase a car.

Implement

Rate of change: $12/hour

Specific point: (25, 2,800)

Best form of line:

Explain

Example 6

Write an equation using the most appropriate form of a line for the given situation. Explain why the graph would be discrete or continuous.

David needed to purchase lumber for a building project. He needed 4 × 4 pieces that cost $8 and 2 × 4 pieces that cost $5. He had $45 to spend on wood.

Implement

14A EXPLORE

> ☑ **Checkpoint**
>
> Laynee is a driver for a food delivery service. She earned $60 after making 15 deliveries and $88 after 22 deliveries. Explain why the graph would be discrete or continuous. Then write a linear equation in the most appropriate form.

PRACTICE 1   14A

Practice 1

Complete the problems on a separate sheet of paper.

Use the domain and range intervals below to choose the graph that best matches the interval, and fill in the blank with the corresponding letter. Then, write the domain and range for each graph in interval notation.

_____ 1) 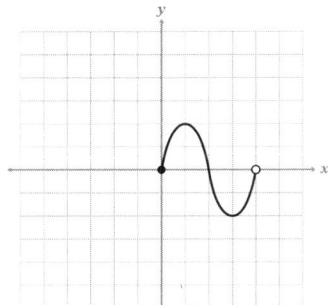   A)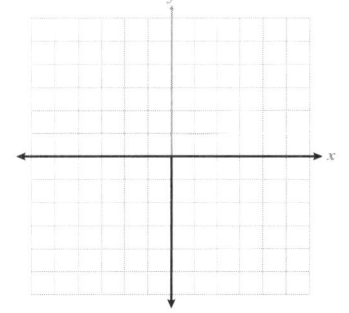

_____ 2) 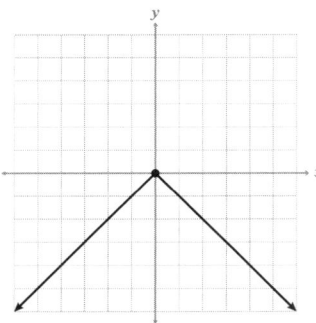   B)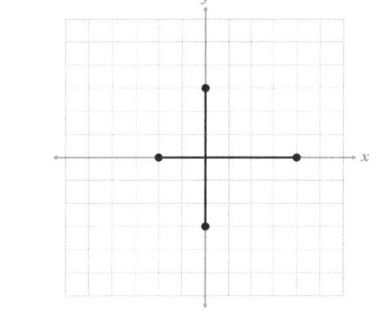

_____ 3) 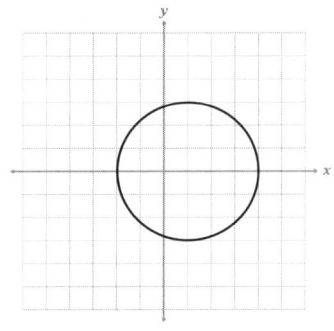   C)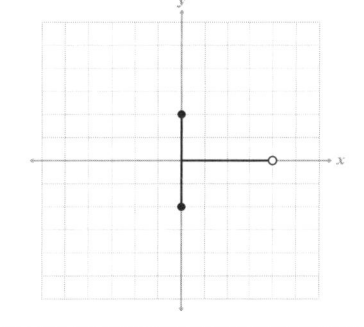

4)   Are all of these graphs functions? Explain.

14A PRACTICE 1

Create a graph from the equation or table of values. Then, write the domain and range for the function in interval notation or set notation.

5) $y = 3x$

6)

x	y
-2	2
-1	1
$-\frac{1}{2}$	$\frac{1}{2}$
0	0
$\frac{1}{2}$	$\frac{1}{2}$
1	1
2	2

Determine if the function is discrete or continuous.

7) $g(x) = 3x - 2$ with a domain of $(-\infty, \infty)$.

8) $h(x) = -x - 4$ with the domain $\{-1, 0, 1, 2, 3\}$.

9) The temperature at a given time of day.

10) The number of sprouted plants in a garden in the month of May.

Write the equation for the line. Choose the best form of the line to write for the given scenarios. Explain how you know the function is discrete or continuous.

11) There were two inches of snow on the ground. Then, during the snowstorm, snow was accumulating at half an inch per hour.

12) Alana was packing boxes to be shipped. Before lunch, she packed 8 boxes in 12 minutes. After lunch, she packed 42 boxes in 28 minutes.

13) The auto repair shop in town charges $175 to analyze the problem for a vehicle. After having his car analyzed, Jimmy was charged $49 per hour to make the repairs.

14) Hector was given $30 this month in spending money. He decided that he would purchase a coffee every day that would cost $3. After how many days and cups of coffee will Hector run out of money? Explain.

15) Grapes cost $4 per pound and apples cost $3 per pound. Kervon had a total of $24 to spend on fruit.

Mastery Check

Show What You Know

You are making an energy snack mix and need to have a total of 33 grams of protein in each snack bag. Bridge Ultra Energy mix has 6 grams of protein per ounce and granola has 3 grams of protein per ounce.

A) Write the equation of the line. Name the form you used.

B) Explain why you believe this is a discrete or continuous function.

C) Graph your equation from Part A. Name the domain and range.

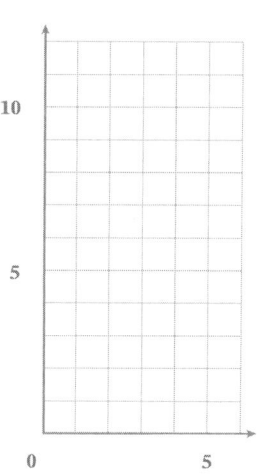

D) Explain the intercepts and how the graph can help determine possible combinations for the snack bags.

Say What You Know

In your own words, talk about what you have learned using the objectives for this part of the lesson and your work on this page.

14A   PRACTICE 2

✏️ Practice 2

Complete the problems on a separate sheet of paper.

Use the domain and range intervals below to choose the graph that best matches the interval, and fill in the blank with the corresponding letter. Then, write the domain and range for each graph in interval notation.

_____ 1)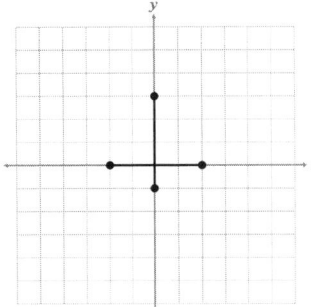

_____ 2)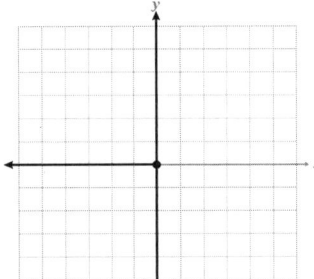

_____ 3)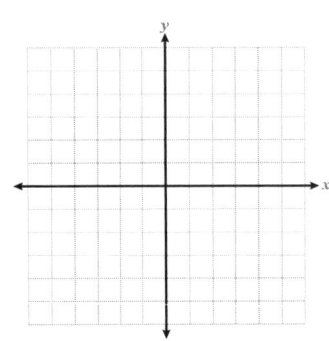

A)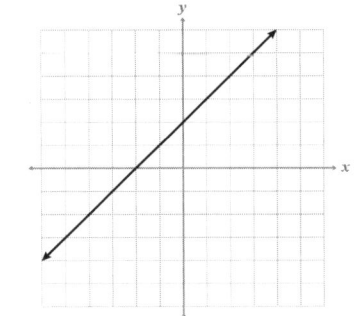

B)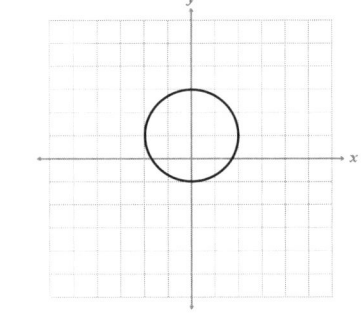

C)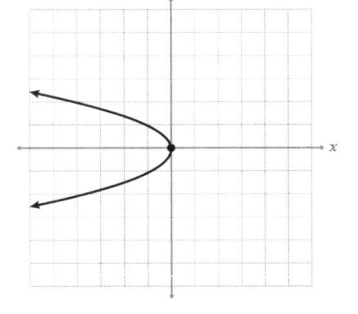

4) Are graphs A, B, and C all functions? Explain.

PRACTICE 2   14A

Create a graph from the equation or table of values. Then, write the domain and range for the function in interval notation or set notation.

5)

x	y
0	0
$\frac{1}{4}$	$\frac{1}{2}$
1	1
4	2

6) $y = x + 2$

Determine if the function is discrete or continuous.

7) $h(x) = -\frac{1}{2}x - 6$ with the domain $\{-2, 0, 2, 4, 6\}$

8) $g(x) = x - 2$ with a domain of $(-\infty, \infty)$.

9) The speed of a car traveling on a local highway.

10) The total number of questions on a worksheet.

Write the equation for the line. Choose the best form of the line to write for the given scenarios.

11) A caterer charges $25 per person plus a deposit of $500.

12) What would the cost be for 40 people?

13) Cardi and Billie were on a road trip and reached an elevation of 14,360 feet. Seventy-five minutes later, their elevation was 3,110 feet above sea level.

14) Explain what the rate of change means in this context.

Brooke could download songs for $1 and movies for $5. She has a total of $24 to spend on media this month.

15) What is the maximum number of movies Brooke is able to purchase? Explain.

16) Would Brooke be able to purchase 3 movies and 9 songs? Show or explain your work.

Siggy and her family rented a car while on vacation. Speedy Rentals charged $17 per day. When Siggy returned the car, she paid $283 for her 9-day rental.

17) Write an equation to model the situation. Is this a discrete or continuous function?

18) What fee did Speedy Rentals charge Siggy? Show or explain your work.

Planet Exercise charges a $10 joiner's fee and then $5 per month for a gym membership.

19) Determine if it is discrete or continuous. Explain your reasoning.

20) Write the equation for the line. Choose the best form of the line to write for the given scenario.

21) Explain why it does not make sense to graph the equation in the second quadrant.

22) If Nicole has a yearly budget of $65, would she be able to afford a membership to Planet Exercise? Explain.

Part B: Arithmetic Sequences

Objectives

In this part of the lesson, you will learn about arithmetic sequences.

By the end of this lesson, you will be able to do the following:

- ⊘ Describe the arithmetic sequence of a given set.
- ⊘ Use a sequence to find additional terms.

Why?

Linear equations all contain patterns, these can also be represented as a sequence of terms. When an arithmetic sequence occurs, it also represents a linear relationship.

Warm Up

1) Assuming a steady pace for each runner, fill in the missing race times.

	1.5 miles	2 miles	2.5 miles	3 miles
Mina		14 min	17.5 min	21 min
Albert	12 min		20 min	24 min
Desiree	9.75 min	13 min	16.25 min	
Toby	18 min	24 min		36 min

2) Assuming that each of the runners maintains the same running rate for all 3 miles, write a sentence describing their rate of change.

Arithmetic Sequences

- A _____ is a discrete function which has a domain of only natural numbers.

- Each term is labeled with a _____ that indicates its position in the sequence.

- _____: a particular type of sequence used in mathematics that contains a particular pattern.

  - Begins with the first term in the sequence, a_1.

  - Adds the same value, d, to each term to get to the next term in the sequence.

    - This value, d, is referred to as the _____ because it is the value that results from subtracting any two consecutive terms.

  - Uses ellipsis (...) to indicate that the sequence has more terms that are not listed and continues on for n-terms.

- The explicit formula for an arithmetic sequence where n is the nth-term being found is:

Example 1

Determine if the following sequences are arithmetic.

A) $\{20, 30, 40, 50, ...\}$
   +10 +10 +10

   $a_1 = 20$
   $d = +10$

   This is arithmetic because the numbers increase by a *constant rate* of 10. Another way of saying a constant rate is the common difference of 10.

B) $\{10, 100, 1000, 10000, 100000, ...\}$

   This is *not* arithmetic because the numbers increase by a *multiple* of 10.

C) $\{10, 5, 0, -5, -10, ...\}$

14B EXPLORE

Example 2

Use sequence G to answer the following questions.

$G = \{3, 7, 11, 15, 19, ...\}$

Plan Identify the rule for this sequence.

What term is a_1?   1 (first)

What is the value of a_1? $a_1 = 3$

What is the value of d? $d = 15 - 11$ (or any two consecutive terms) $= 4$

Rule in words:

What term is a_3?

What is the value of a_3?

What is the value of a_5?

What is the value of a_6?

Using the explicit formula, $a_n = a_1 + d(n-1)$, find a_8 and a_{21}.

$a_n = a_1 + d(n-1)$

$\qquad\qquad\qquad\qquad$ Rewrite the formula with a_1 and d.

$n = 8 \qquad\qquad\qquad n = 21$

$a_n = 3 + 4(n-1)$

$a_8 = 3 + 4(8-1)$

$a_8 = 3 + 4(7)$

$a_8 = 31$

> ☑ **Checkpoint**
>
> Given the following sequence, find a_1 and d.
> Then, use the formula $a_n = a_1 + d(n-1)$ to solve for a_3 and a_{32}.
>
> $A = \{4, 7, 10, 13, ...\}$

Practice 1

Complete the problems on a separate sheet of paper.

Determine if the following sequences are arithmetic. If so, find the common difference and the next three terms in the sequence (a_5, a_6, a_7).

1) $\{2, 4, 8, 16, ...\}$
2) $\{2, 4, 6, 8, ...\}$
3) $\{6, 5.5, 5, 4.5, ...\}$
4) $\{20, 10, 5, 2.5, ...\}$

Use the arithmetic sequence formula, $a_n = a_1 + d(n-1)$.

5) $\{2, -1, -4, -7, -10, ...\}$
   What is the value of a_1?
   What is the value of d?
   Use the formula to find a_5, a_{10}, and a_{12}.

6) $\{-5, -4\frac{2}{3}, -4\frac{1}{3}, -4, ...\}$
   What is the value of a_1?
   What is the value of d?
   Use the formula to find a_5, a_{10}, and a_{12}.

When a sequence is graphed, the domain values represent the terms of the sequence $\{1, 2, 3, ..., n\}$. The range is represented by the value of a_n. Another way to think about this is (n, a_n). The graph below represents an arithmetic sequence.

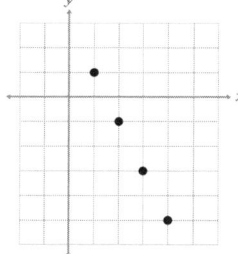

7) List the first 4 values of the range.
8) List the first 4 values of the domain.
9) Write the formula for the sequence for the nth term.
10) Write the arithmetic sequence equation found in problem 9, in slope-intercept form and function notation.

---

11) The common difference of an arithmetic sequence is $\frac{1}{3}$. If $a_7 = 5$, what is a_1? Show your work.

12) The common difference in an arithmetic sequence is -4. If $a_{11} = -2$, what is a_1?

Aria and Reza were trying to go on as many rides as possible at the amusement park. After 1 hour, they had been on 2 rides, in 1.5 hours they had been on 5 rides, and in 2 hours they had been on 8 rides.

13) Write the arithmetic sequence supposing this continues until the park closes.
14) Write the formula for the sequence.
15) Suppose the park was open for 12 hours, how many rides did Aria and Reza go on?

Eliza's job requires her to be able to lift boxes that weigh up to 40 pounds. She is moving boxes in the warehouse and wants to track the total amount she lifted in one day. Below is the table Eliza used to track her totals for the day.

Box #	1	2	3	4	5
Total weight	32	64	96	128	160

16) Write the formula for the arithmetic sequence.
17) How many pounds will Eliza have moved for a_{15}? Show your work.

14B MASTERY CHECK

Mastery Check

Show What You Know

Kalen was stacking chairs in the school gym after a music concert. Starting with one chair, the stack was 18 inches off the ground. With two chairs, the stack was 23 inches off the ground, and with three chairs, it was 28 inches.

A) Write a formula to find the nth number of chairs in a stack.

B) If the maximum height allowed for stacking chairs is 58 inches, how many chairs can be put in a stack?

C) Write the domain and range for this problem. Why does this sequence have an ending point?

D) Write the equation in slope-intercept form. What do the variables x and y represent?

E) Why does the y-intercept *not* make sense for the context of this problem?

Say What You Know

In your own words, talk about what you have learned using the objectives for this part of the lesson and your work on this page.

Practice 2

Complete the problems on a separate sheet of paper.

Determine if the following sequences are arithmetic. If so, find the common difference and the next three terms in the sequence (a_5, a_6, a_7).

1) $\{1, 4, 9, 16, ...\}$

2) $\{3, 9, 27, 81, ...\}$

3) $\{-1, -2, -3, -4, ...\}$

4) $\{20, 15, 10, 5, ...\}$

Use the formula for an arithmetic sequence, $a_n = a_1 + d(n-1)$, to find a_5, a_{10}, and a_{12}

5) $d = -\frac{1}{2}, a_1 = 26$

6) $d = 4, a_1 = -5$

7) The first four houses on Green Street are numbered 201, 208, 215, and 222. What house (term) is numbered 264?

8) Rolanda drives 22 miles each day. Today her odometer read 13,426 miles. Write a formula for the nth term, then solve for the odometer reading on day 45.

9) The common difference for an arithmetic sequence is -0.25. If $a_{21} = 6$, what is a_1?

Mina Rees wanted to determine a formula for the following graphed arithmetic sequence.

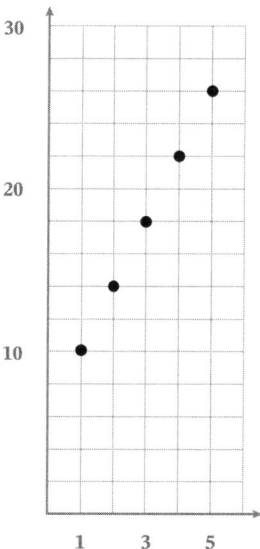

10) Find the formula for the arithmetic sequence.

11) Write an equation in slope-intercept form to represent the graph.

12) What do the common difference, d, and the rate of change, m, have in common?

Create your own arithmetic sequence with a common difference of $\frac{1}{5}$.

13) List the first 5 terms of your sequence.

14) Write the formula for the nth term.

15) Find a_9 using your formula.

TARGETED REVIEW 14

◎ Targeted Review

In the Targeted Review, you will practice topics you have mastered in earlier lessons. Reviewing these concepts will help you be successful as you work through this unit.

Complete the problems on a separate sheet of paper.

1) Write the equation in slope-intercept form given the points $(-4, -2)$ and $(-3, 1)$.

2) Convert the equation for the line from standard form to slope-intercept form: $2x - 3y = -6$

3) Rook and Tripp painted pictures. The equations below represent how many pictures they had ready for sale after a given number of hours. Determine who painted faster and how much faster he painted than the other.

$$\text{Rook: } p = 4h + 5 \qquad \text{Tripp: } p = 12h - 2$$

4) Write the equation for the horizontal line and vertical line that passes through the point $(4, 5)$.

5) Is relation M a function? Explain your answer.
$M: \{(1, 3), (4, -2), (9, 1), (3, 4), (5, 3)\}$

6) Write the equation perpendicular to the given line, $y = -\frac{1}{4}x + 11$, through $(-3, 8)$.

7) Jimena used the equation $r(w) = -25w + 500$ to represent her lunch budget for her first semester at college, where w represents weeks and $r(w)$ represents the remaining money after w weeks. Explain the meaning of the slope and y-intercept related to the problem.

8) Write a linear equation. Hector saved $20 each week from his paycheck. After working for 6 weeks, he had $520 in his savings account. Write a linear equation to model this situation. Name the form of the equation used.

Multiple Choice

_____ 9) Determine the correlation coefficient that best matches the scatter plot and trend line.

A) $r = -0.91$
B) $r = -0.42$
C) $r = 0.43$
D) $r = 0.91$

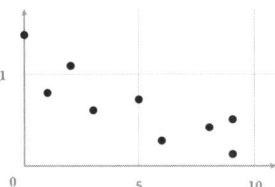

_____ 10) Which point lies on the line $2x + y = 1$?

A) $(2, -3)$
B) $(2, 1)$
C) $(2, 0)$
D) $(2, 3)$

_____ 11) Choose the equation that best matches the given scatter plot.

A) $y = -0.1x + 1.15$
B) $y = -10x + 1.15$
C) $y = 0.1x + 1.15$
D) $y = 10x + 15$

_____ 12) Given the table, write the equation in slope-intercept form.

A) $y = \frac{4}{3}x$
B) $y = \frac{3}{4}x$
C) $y = -\frac{4}{3}x$
D) $y = -\frac{3}{4}x$

x	$z(x)$
-4	-3
0	0
4	3
8	6

UNIT 3 Record Keeping

Name: _____ Algebra 1

Place a checkmark or the date when items are completed. For tests, record the score if preferred.

Lesson		Part	Guided Notes	Practice 1	Practice 2	Mastery Check	Targeted Review	Lesson Test
15 Graphing Systems	A	Graphing Systems of Equations						
	B	Graphing Systems of Inequalities						
16 Solving Systems of Equations Algebraically	A	Solving Systems Using Substitution						
	B	Solving Systems Using Elimination						
17 Applications of Linear Systems	A	Efficiently Solving Systems						
	B	Applications of Linear Systems						
18 More Applications of Linear Systems	A	More Applications of Linear Systems						
	B	Applications of Linear Inequalities						

Unit 3 Test Date _____ Score _____

Midterm Test Date _____ Score _____

Lesson Objectives

Confirm with your instructor and check each objective that you have mastered.

Lesson 15: Part A
- ☐ Identify and describe the types of possible solutions for a system of equations.
- ☐ Graph a system of equations given in slope-intercept form. Then find and explain the solution to the system.
- ☐ Graph a system of equations not given in slope-intercept form. Then find and explain the solution to the system.

Lesson 15: Part B
- ☐ Describe what the shaded region on a graph of a linear inequality or a system of linear inequalities represents.
- ☐ Graph a linear inequality.
- ☐ Graph a system of linear inequalities.

Lesson 16: Part A
- ☐ Use substitution to solve a system of equations when both equations have an isolated variable.
- ☐ Use substitution to solve a system of equations when one equation has an isolated variable.
- ☐ Use substitution to solve a system of equations when none of the equations have an isolated variable.

Lesson 16: Part B
- ☐ Use elimination to solve a system of equations when one variable has a set of opposite coefficients.
- ☐ Use elimination to solve a system of equations by distributing −1 across an equation.
- ☐ Use elimination to solve a system of equations using linear combinations.

Lesson 17: Part A
- ☐ Choose the best method to solve a system of linear equations and justify your choice.

Lesson 17: Part B
- ☐ Write systems of equations for how much and how many, coin, and wind and water word problems and solve them.
- ☐ Explain the solutions to how much and how many, coin, and wind and water problems and determine if the solutions are reasonable.

Lesson 18: Part A
- ☐ Write systems of equations for break-even, formula, and digit word problems and solve them.
- ☐ Explain the solutions to break-even, formula, and digit problems and determine if the solutions are reasonable.

Lesson 18: Part B
- ☐ Identify and apply the following key words to write an inequality from a verbal model:
  - is more than
  - is less than
  - is no more than
  - is at least
  - is at most
  - has a minimum of
  - has a maximum of
- ☐ Write a system of inequalities from a scenario and solve for the solution.
- ☐ Explain what the solution to a system of inequalities means within a given context.

Lesson 15
Graphing Systems of Linear Equations and Inequalities

Outline

Part A
Graphing Systems of Equations

- Systems and Their Solutions
- Systems of Equations in Slope-Intercept Form
- Systems in a Variety of Forms

Part B
Graphing Systems of Inequalities

- Solutions to Linear Inequalities
- Graphing Linear Inequalities
- Graphing Systems of Linear Inequalities

Targeted Review

Vocabulary

- system of equations
- solution (to a system of equations)
- coincident lines
- solutions (to an inequality)
- system of linear inequalities

15A EXPLORE

Part A: Graphing Systems of Equations

Objectives

In this part of the lesson, you will learn about graphing systems of equations.

By the end of this lesson, you will be able to do the following:

- Identify and describe the types of possible solutions for a system of equations.
- Graph a system of equations given in slope-intercept form. Then find and explain the solution to the system.
- Graph a system of equations not given in slope-intercept form. Then find and explain the solution to the system.

Why?

What do businesses, engineers, and air traffic controllers have in common? They all use systems of equations. From predicting profits, planning and building cities, or charting safe flight paths, many diverse careers use systems of equations to solve problems.

Warm Up

1) Write the equations of the two horizontal lines.

2) Write the equations of the two vertical lines.

3) Name all the points where the lines intersect.

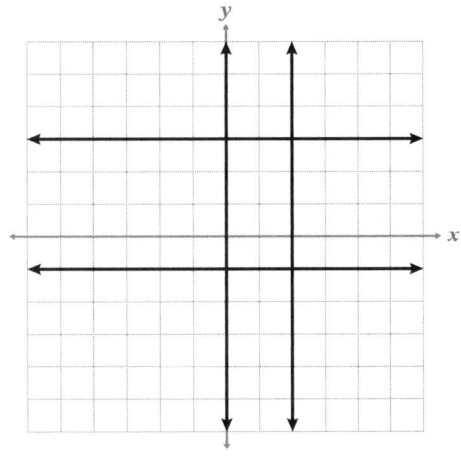

Systems and Their Solutions

- A _____ is two or more equations grouped together.

- The _____ to a system of linear equations is an ordered pair where the lines of the system intersect.

- Linear systems of equations have three types of solutions:

EXPLORE 15A

Example 1

Name the number of solutions for the graphed systems of equations.

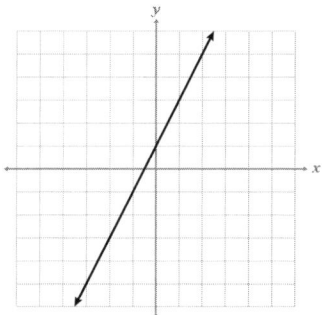 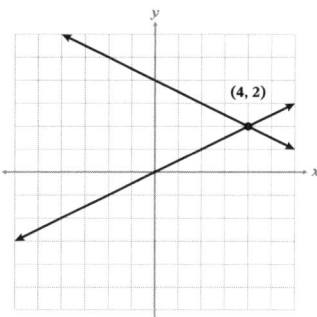 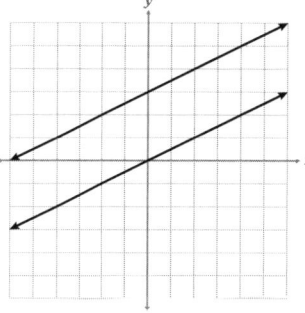

_____   _____   _____

- The graphed equations for systems of linear equations that have one solution will intersect at exactly

  _____ on the coordinate plane.

- A system with one solution occurs when lines have _____ slopes.

- Both equations have (4, 2) as a solution; this is the _____

  to the system. If the ordered pair does not make both equations true, then it is *not* a solution.

- A system with zero solutions is a system of _____ lines.

- _____ are lines that are graphed exactly on top of

  one another.

- The solution to a system of coincident lines is any point on the line or infinite solutions

  _____ .

☑ Checkpoint

Write the equations of the lines in the system from the given graph. Then find the solution.

Solution:

System:

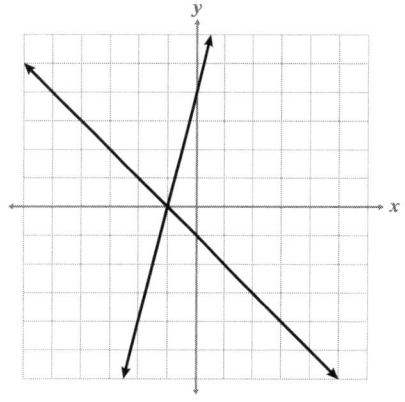

15A EXPLORE

Systems of Equations in Slope-Intercept Form

- When graphing a system of equations on a coordinate plane, graph the _____ across the _____ so that you can find the intersection point.

Example 2
Given the system of equations, draw the graph and identify the solution.

$y = -\frac{3}{4}x + 1$

$y = \frac{1}{2}x + 6$

Plan Graph both lines on the same coordinate plane until the solution is found.
Check the solution algebraically.

Implement

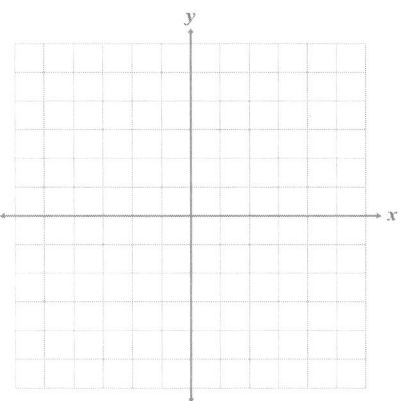

✓ Checkpoint

Graph the system of equations and name the solution.

$y = 2x - 1$
$y = -3x + 4$

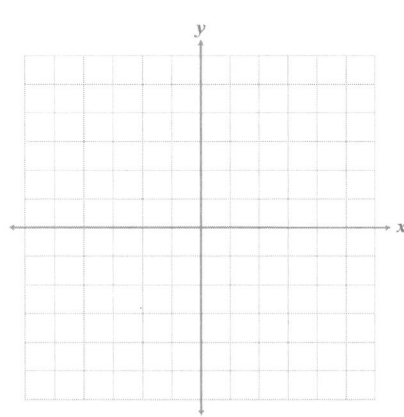

Solution:

Systems in a Variety of Forms

- When the equations in a system of equations are not in the same form, choose _____ and rewrite the equations before graphing, or graph each equation from its given form.

- The _____ way to graph equations is to leave them in their given form.

Example 3

Graph the system of equations and name the solution.

$x - y = 6$
$x + 3y = -2$

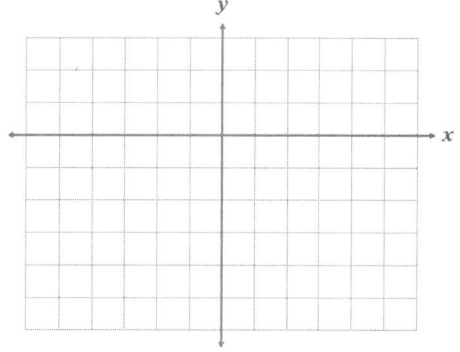

Plan Find the x- and y-intercepts
Graph the lines
Find the solution

Implement

Equation 1: $x - y = 6$ Equation 2: $x + 3y = -2$
$(x, 0)$ $(x, 0)$
$x - 0 = 6$

Check
$x - y = 6$ $x + 3y = -2$

$(0, y)$ $(0, y)$
$0 - y = 6$
$-y = 6$

$$m = -\left(\frac{A}{B}\right)$$

15A EXPLORE

Example 4

Graph the system of equations and name the solution.

$$3x + 5y = 0$$
$$y + 7 = 2(x - 3)$$

Equation 1: $3x + 5y = 0$ Equation 2: $y + 7 = 2(x - 3)$

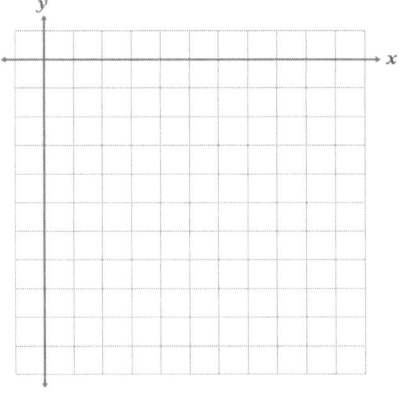

Check

☑ Checkpoint

Graph the system of equations and name the solution.

$$y = \tfrac{1}{3}x + 3$$
$$y - 3 = \tfrac{1}{2}(x + 2)$$

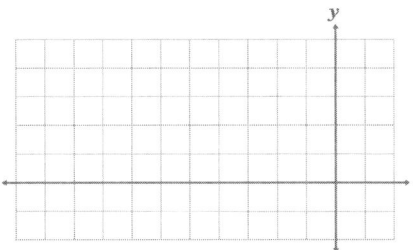

Practice 1

Complete the problems on a separate sheet of paper.

Fill in the blanks with the correct vocabulary word.

1) Lines that have equal slope are either coincident lines or _____ lines.

2) Systems of linear equations can have _____, one, or _____ solutions.

Determine if (3, −1) is a solution to the given systems. Show your work.

3) $x - y = 4$
   $x + y = 3$

4) $2x + 5y = 1$
   $y = \frac{1}{2}x - \frac{1}{4}$

Use the graph of lines a, b, and c to find the solution between the given equations.

5) line a and line b

6) line c and line b

Given the graph of lines p and q, write the equations of the system and find the solution.

7) line p and line q

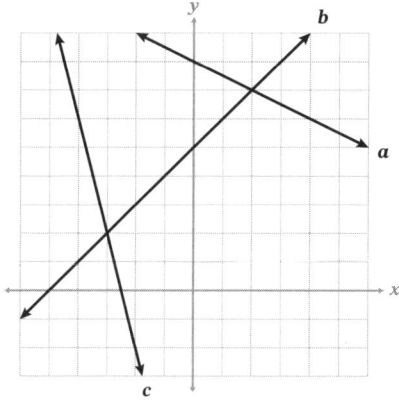

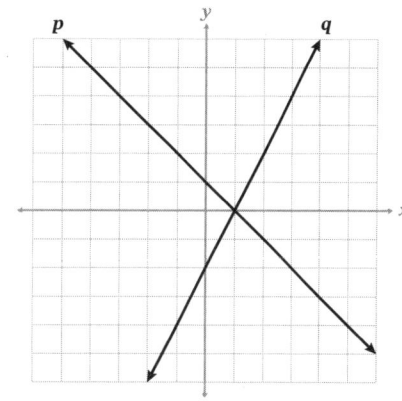

Draw a sketch of the three types of solutions to a system of equations on a separate sheet of paper.

8) zero solutions

9) one solution

10) infinite solutions

Graph the system of equations. Mark and name the solution.

11) $y = -2x + 5$
    $y = \frac{5}{2}x + 5$

12) $y = x$
    $y = -2x - 3$

13) $y = \frac{1}{2}x - 1$
    $y - 4 = \frac{1}{2}(x - 2)$

14) $y + 5 = x + 4$
    $2x - 3y = 5$

15) $2x + y = 8$
    $y = x + 5$

16) $x - y = 2$
    $y - 2 = x - 4$

15A MASTERY CHECK

Mastery Check

Show What You Know

Airplane flights between Historyville and Mathtopia run multiple times each day. There is an outgoing flight path as well as an incoming flight path between these two cities. The equations below represent the runways for the airport in Mathtopia.

Outgoing flights: $y = -\frac{1}{2}x + 2$ Incoming flights: $x + 2y = 10$

A) Graph the outgoing and incoming runways as lines on the given coordinate plane. Label the lines as outgoing and incoming.

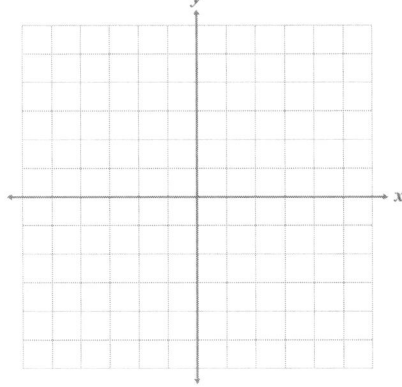

A) Describe the relationship between the runways. Why is this important for the airplanes' flight paths?

B) An access road crosses both runways so that airport employees can enter the runway with baggage to load onto the planes. Graph this equation on the same coordinate plane.

Access Road: $y + 2 = x + 1$

C) Where does the access road intersect each runway?

Outgoing runway: _____   Incoming runway: _____

Say What You Know

In your own words, talk about what you have learned using the objectives for this part of the lesson and your work on this page.

Practice 2

Complete the problems on a separate sheet of paper.

Determine if $\left(-\frac{1}{3}, 4\right)$ is a solution to the given systems. Show your work.

1) Is $\left(-\frac{1}{3}, 4\right)$ a solution for $3x - 2y = -9$ and $3x - 5y = -21$?

2) Is $\left(-\frac{1}{3}, 4\right)$ a solution for $3x + 3y = 11$ and $6x + y = 7$?

Write the system of equations given the graph. Then find the solution.

3)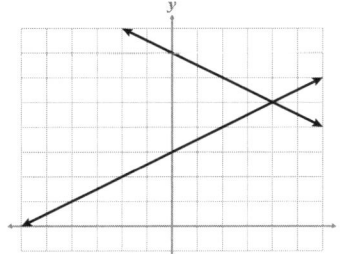

4)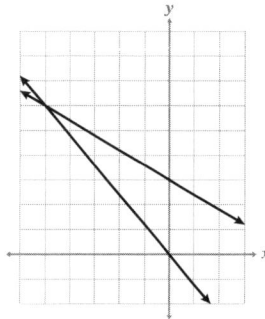

Match the solution to the system of equations.

_____ 5) $2x - 5y = -3$
$2x + 3y = 5$

_____ 6) $2x + 5y = 3$
$3x - 2y = -5$

_____ 7) $x + 3y = -5$
$4x - 3y = -5$

A) $(-1, 1)$
B) $(-2, -1)$
C) $(1, 1)$

Write the system of equations given the graph. Then find the solution.

8)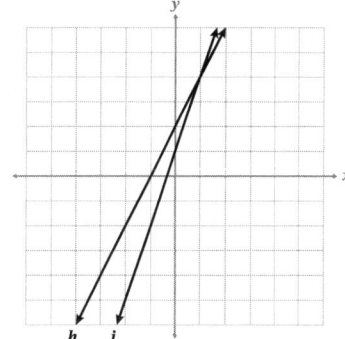

Graph the system of equations. Mark and name the solution.

9) $2x + y = 8$
$y = 2x - 4$

10) $3x + y = 2$
$3x + y = -2$

11) $2x - y = 1$
$x - y = -1$

12) $y = 3$
$x = -2$

13) $x = 2 - 2y$
$2x = -2 - y$

14) $3y + 2x - 3 = 0$
$6 + 2x + 3y = 0$

15B EXPLORE

Part B: Graphing Systems of Inequalities

Objectives

In this part of the lesson, you will learn about graphing systems of inequalities.

By the end of this lesson, you will be able to do the following:

- Describe what the shaded region on a graph of a linear inequality or a system of linear inequalities represents.
- Graph a linear inequality.
- Graph a system of linear inequalities.

Why?

Being able to read, explain, and create graphs of systems of inequalities will help you better understand how inequalities, equations, and their graphs all relate. This will help you master future topics in algebra.

Warm Up

Solve the inequalities. Graph the solution on a number line.

1) $-\frac{2}{5}x + 7 > 1$

2) $0 < 3y - 11 < 3$

3) What does the shaded portion of your number line represent?

Solutions to Linear Inequalities

- The following rules apply to linear inequalities:

  - The _____ symbol is replaced by one of the inequality symbols ($<$, $>$, $\geq$, or $\leq$).

  - A region of the coordinate plane is _____.

  - The inequality will be in one of the _____ forms.

- The graph of a linear inequality is shaded to show _____

  that will make the inequality _____.

EXPLORE 15B

Four Types of Linear Inequalities

Symbol	Wording	Represented Graphically
>	is greater than	
<	is less than	
≥	is greater than or equal to	
≤	is less than or equal to	

Example 1

Given the graph of the linear inequality, describe the solution.

$y \geq 2x + 1$

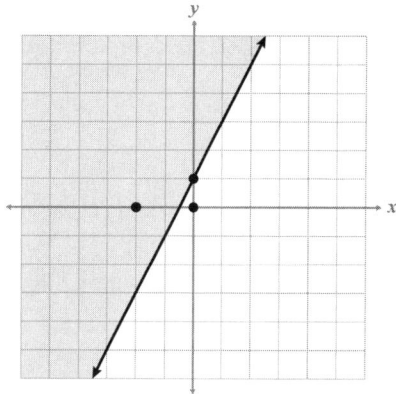

First, identify familiar values in the inequality.

Pick test points above, on, and below the line to check the inequality algebraically.

Above the line: $(-2, 0)$

$y \geq 2x + 1$

$0 \geq 2(-2) + 1$

$0 \geq -4 + 1$

$0 \geq -3$

On the line: $(0, 1)$

$y \geq 2x + 1$

$1 \geq 2(0) + 1$

$1 \geq 1$

Below the line: $(0, 0)$

$y \geq 2x + 1$

$0 \geq 2(0) + 1$

$0 \geq 0 + 1$

$0 \geq 1$

The solution to the inequality is *all* ordered pairs *on* and *above* the graphed line.

Example 2

Identify key features of the given graph, then write the linear inequality.

line:

shading:

Mark the following on the graph: m, b

linear inequality:

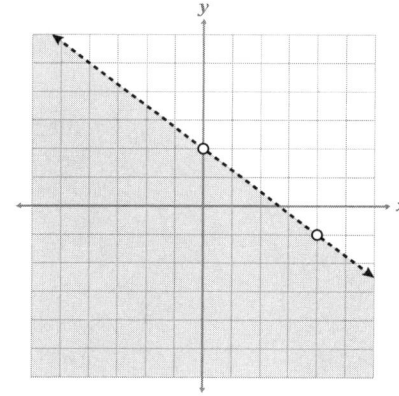

15B EXPLORE

☑ Checkpoint

Identify key features of the given graph, then write the linear inequality.

line: solid or dashed
shading: above or below the line

inequality symbol:

$m =$ $\qquad$ $b =$

inequality:

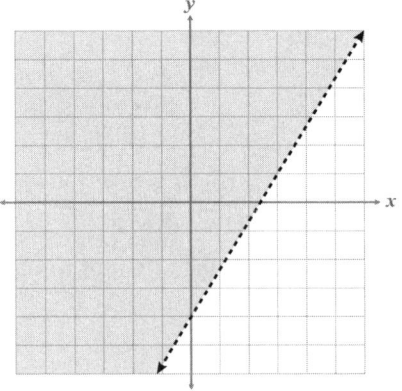

▶ Graphing Linear Inequalities

■ The key difference between graphing linear inequalities and equations is determining if the line will be _____ or _____ and then _____ the region where the solutions lie.

Example 3

Graph the inequality: $3x - 2y \leq 8$

Plan Write in slope-intercept form.
Identify characteristics of the graph.
Graph the inequality.

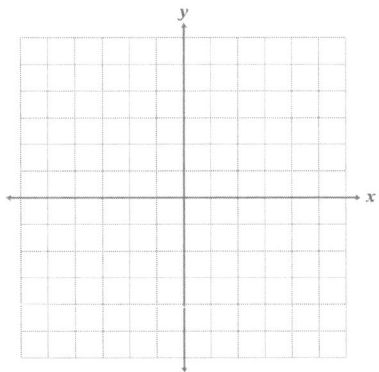

Implement

Identify the graph characteristics
line: solid or dashed

shading: above or below the line

Explain

◀ The line is solid because there is an =.

◀ The shading is above the line because the inequality switches directions, and greater numbers are above the line.

◀ The y-intercept is $(0, -4)$ and a closed point.
The slope is $\frac{3}{2}$.

Example 4

Graph the linear inequality: $y < 3x - 1$

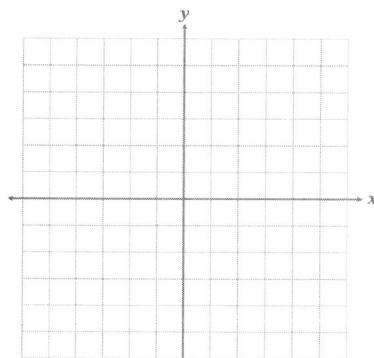

Identify the graph characteristics

line: solid or dashed

shading: above or below the line

open or closed point

Explain

◀ The line is dashed because there is no "=."

◀ The shading is below the line because numbers with less value are negative.

◀ The y-intercept is $(0, -1)$ and marked with an open point. The slope is $\frac{3}{1}$.

☑ Checkpoint

Identify the key information for the linear inequality, then graph it.

$y < \frac{1}{4}x + 1$

line: solid or dashed

shading: above or below the line

$m =$ $b =$

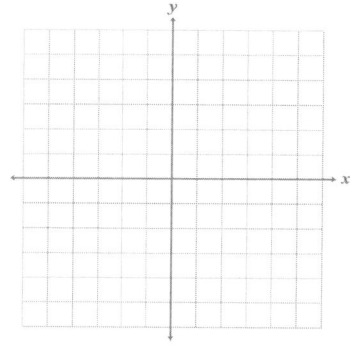

▶ Graphing Systems of Linear Inequalities

- A _____ is two or more linear inequalities grouped together.

- Graphing a system of linear inequalities follows the same rules as graphing one _____.

- The biggest difference is the solutions to a system of linear inequalities are in the shaded _____ of the graph where the linear inequalities _____.

15B EXPLORE

Example 5

Graph the system of inequalities.

$y \leq -\frac{1}{2}x + 1$

$y < x$

Plan Identify the key information for each inequality.
Graph each inequality.

Implement

Inequality 1: $y \leq -\frac{1}{2}x + 1$

line: solid

shading: below

$m = -\frac{1}{2}, b = 1$, closed point

Inequality 2: $y < x$

line: _____

shading: _____

$m = $ _____, $b = $ _____, _____ point

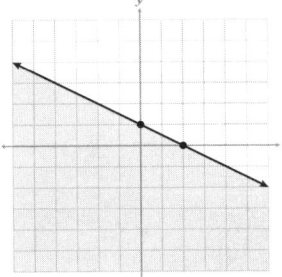
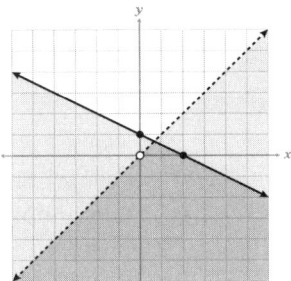

The portion of the graph where the shaded regions overlap is the solution to the system of inequalities.

Example 6

Graph the system of inequalities. Name the quadrant(s).

$y < 2x + 3$
$2x - y < 1$

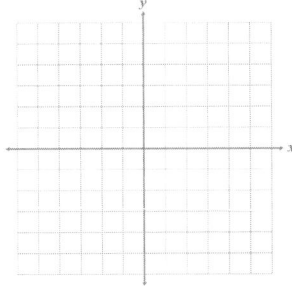

☑ Checkpoint

Graph the system of inequalities.

$y < 5x - 2$

$y > \frac{2}{3}x$

Name the quadrant or quadrants where the solution is located.

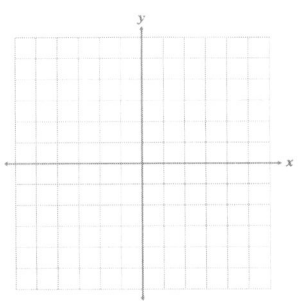

Practice 1

Complete the problems on a separate sheet of paper.

1) Given the graph, identify the slope, y-intercept, if the line is solid or dashed, and if the shading is above or below the line.

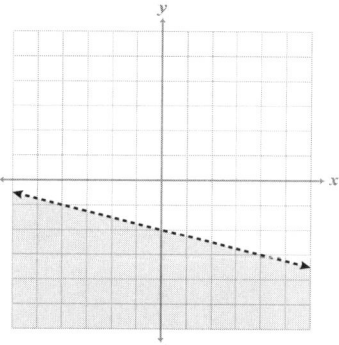

2) Given $y \geq x$, identify the slope, y-intercept, if the line is solid or dashed, and if the shading is above or below the line. Then graph.

Write the sentences using one of the following: always, sometimes, never. (Answer choices may be used more than once.)

3) Points on the line are _____ solutions to a linear inequality.

4) The shaded region _____ represents all possible solutions.

5) When dividing or multiplying by a negative coefficient, the inequality symbol _____ changes directions.

6) Graph $y \leq -\frac{2}{3}x + 4$

7) Write the system of inequalities given the graph.

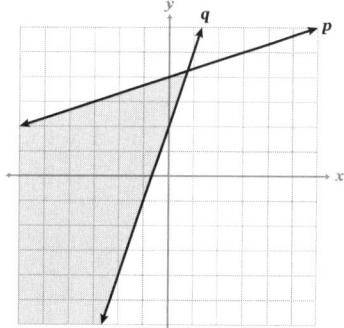

Use the inequalities in problem 7 to prove if the following ordered pairs are solutions.

8) $(-2, -4)$

9) $(4, 1)$

15B   PRACTICE 1

10) Write the system of inequalities given the graph.

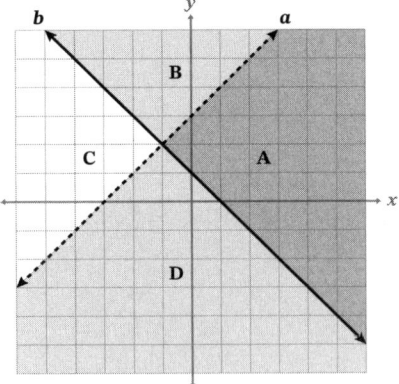

From the graph, determine which region (A, B, C, or D) each point is in and whether it is a solution to the system.

11) (2, 0)

12) (−2, −2)

13) (0, −3)

14) (5, −2)

15) (−3, 4)

---

16) Identify the slope, y-intercept, if the line is solid or dashed, and if the shading is above or below the line. Then graph.

Inequality c: $y \geq 3x - 3$ Inequality d: $y \geq \frac{1}{2}x + 2$

	Inequality c	Inequality d
line		
shading		
b		
m		

Graph the system of inequalities.

17) $y \leq \frac{3}{4}x + 4$
$2x + 3y \leq 4$

18) $y \leq -\frac{1}{3}x - 2$
$y > 3x - 2$

19) $3x + 4y < 12$
$y < -\frac{3}{2}x - 3$

MASTERY CHECK 15B

Mastery Check

Show What You Know

A) Graph the system of inequalities:
$y < 2$
$x > -1$
$y \geq x$

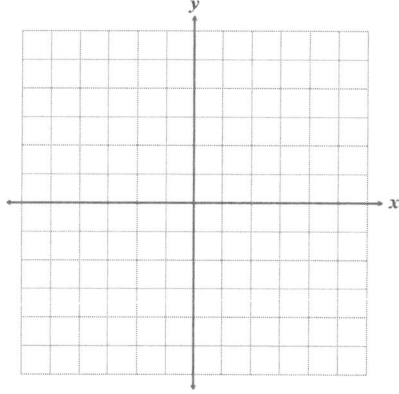

B) What is the shape formed by the three inequalities on the coordinate plane?

C) Mark where the lines in part A intersect. Name the three vertices of the figure formed on the coordinate plane.

D) Find the area of the figure using a formula from your Formula Sheet. Write the formula, then solve.

Say What You Know

In your own words, talk about what you have learned using the objectives for this part of the lesson and your work on this page.

Practice 2

Complete the problems on a separate sheet of paper.

1) What does the shaded region represent for a linear inequality on the coordinate plane?

2) Given the graph, identify the slope, y-intercept, if the line is solid or dashed, and if the shading is above or below the line.

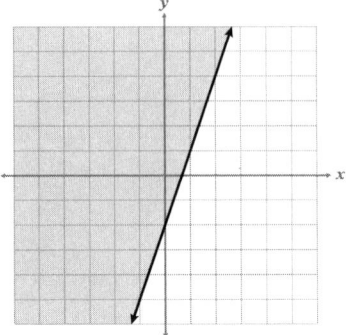

Graph.

3) $y > \frac{7}{4}x - 3$

4) $2x - y > 3$

5) $y > \frac{5}{3}x - 4$

6) Write the system of inequalities given the graph.

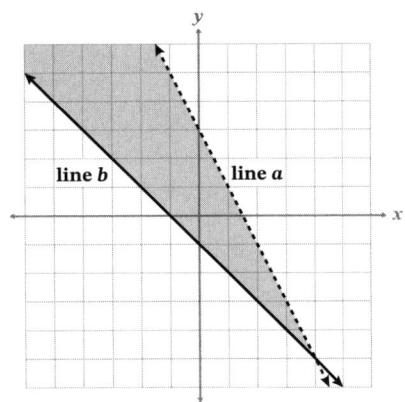

7) Explain how you determined which inequality symbol to use for each linear inequality in problem 6.

8) Identify the slope, y-intercept, if the line is solid or dashed, and if the shading is above or below the line. Then graph.
Inequality m: $5x + 3y \geq -15$
Inequality n: $2x - 3y < 6$

Graph the system of inequalities.

9) $y < 3$
   $x \geq 2$

10) $y < 3x$
    $y > 2x$

11) $y \geq 4x - 2$
    $x + y \geq 0$

12) $y > \frac{5}{4}x - 4$
    $y > -\frac{2}{3}x + 5$

Targeted Review

In the Targeted Review, you will practice topics you have mastered in earlier lessons. Reviewing these concepts will help you be successful as you work through this unit.

Complete the problems on a separate sheet of paper.

1) Write the equation of the line in slope-intercept form using the graph.

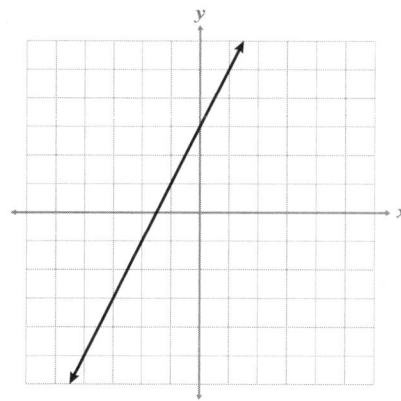

2) Write the equation in slope-intercept form: $5x - 11y = 28$

3) Solve the compound inequality. Graph the solution on a number line.
$6 - 3x \leq -4$ or $2 - x > 4$

Hillary hired someone to paint the outside of her house. The paint cost Hilary $78, and the painter charged $50 per hour to paint the house.

4) Write an equation to find the total cost of painting the house for any number of hours.

5) If the total bill was $253, how long did the painter work?

6) Solve: $\frac{2}{3}(5x - 2) = \frac{1}{4}(x + 8)$

7) Write an equation for the line that is perpendicular to $y = \frac{5}{8}x - 2$ and passes through $(-1, 6)$.

8) Mary-Jo was planning her graduation party at a hall. She was told that 40 people would cost $600 and 54 people would cost $768. What formula is the caterer using to give Mary-Jo the prices? Explain what the slope and y-intercept mean in the context of the problem.

9) Guy's Grocery recorded the number of customers they had over five days. Then Guy did some calculations and determined the following statistics.

Day 1: 1,000	Day 2: 870	Day 3: 900	Day 4: 100	Day 5: 910
mean = 756	mode = none	median = 900		range = 900

What is the best measure to determine the typical number of customers? Explain.

10) Solve the compound inequality:
$\frac{5}{3} < \frac{1}{5}x - 3 \leq 9$

11) Find the x- and y-intercepts as ordered pairs:
$15x + 13y = -30$

Multiple Choice

12) Select the two sentences that are true based on the scatterplot.

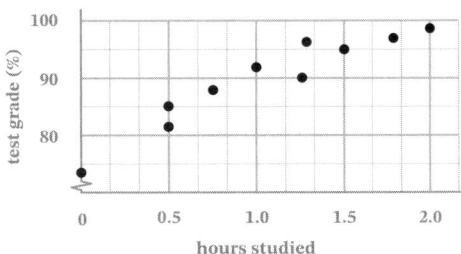

☐ As study time decreases, test grades decrease.

☐ All students scored above 85%.

☐ The exact average for the class was 85%.

☐ Students who studied more earned higher scores.

_____ **13)** Kara documented propane delivery and usage for her family. She tracked the number of gallons the family used monthly and knew that their propane tank held 900 gallons of propane when full. The equation of the line of best fit is shown below.

$$y = -75.8x + 900$$

Which statement correctly describes the slope of the equation in the context of the situation?

A) Each month, Kara spent $900.

E) Each month, Kara's family used about 75.8 gallons of propane.

B) With 900 gallons of propane, Kara's family spends $75.80.

C) For every 75.8 gallons used, the family buys 900 more gallons.

Lesson 16
Solving Systems of Equations Algebraically

Outline

Part A
Solving Systems Using Substitution

- Substitution when Both Equations Have an Isolated Variable
- Substitution: One Variable Isolated
- Substitution: No Variables Isolated

Part B
Solving Systems Using Elimination

- Elimination: Opposite Coefficients
- Elimination: Multiplying by −1
- Elimination: Linear Combinations

Targeted Review

Vocabulary

- linear combinations

16A EXPLORE

Part A: Solving Systems Using Substitution

Objectives

In this part of the lesson, you will learn about solving systems using substitution.

By the end of this lesson, you will be able to:

- ✓ Use substitution to solve a system of equations when both equations have an isolated variable.
- ✓ Use substitution to solve a system of equations when one equation has an isolated variable.
- ✓ Use substitution to solve a system of equations when none of the equations have an isolated variable.

Why?

Knowing how to solve systems of equations using substitution helps you make connections in algebra. When you use substitution, you can easily see those connections.

Warm Up

Determine if the point is a solution for the given equation.

1) Is $(-1, 5)$ a solution for $y = 3x + 2$? Explain.

2) Is $(2.5, -6)$ a solution for $2x + 3y = -13$? Explain.

▶ Substitution when Both Equations Have an Isolated Variable

- To solve a system using substitution you must have an _____.

- The isolated variable is set equal to an expression. You can use this expression to _____ the variable anywhere it appears in the second equation.

EXPLORE 16A

Example 1

This picture equation can be written in symbols using *c* for circles and *t* for triangles.

$c = 2t + 3$

$c + t = 9$

substitution

$3t + 3 = 9$

$3t = 6$

$t = 2$

$c = 2t + 3$ or $c = 2(2) + 3$

$c = 7$

Solving using substitution follows these general steps:

1) Use an _____ variable from one equation to substitute an expression into the remaining equation in the system.

2) Solve the resulting equation for the one _____ variable.

3) Use the value of the variable found to make a second substitution into _____ of the original equations to find the value of the second variable.

4) Write the solution to your system as an _____.

16A EXPLORE

Example 2

Solve the system of equations using substitution.

$y = 4x - 5$
$y = -2x - 2$

Plan In a system of equations, the value of the variables must be the same.
$y = y$ (Reflexive Property)
The equations can equal one another using substitution.
Solve for y.
Solve for x.

Implement   **Explain**

$4x - 5 = -2x - 2$ ◂ Substitution
$+ 2x \quad\quad + 2x$ ◂ Addition Property of Equality
$\quad\quad + 5 \quad\quad + 5$ ◂ Addition Property of Equality
$\quad\quad \frac{6x}{6} = \frac{3}{6}$ ◂ Multiplication Property of Equality
$\quad\quad x = \frac{1}{2}$

Equation 1   **Equation 2**

$y = 4x - 5$

Since the same value was found by checking the equation, the solution is _____.

☑ Checkpoint

Solve the system of equations using substitution.

$y = -x - 13$
$y = x - 5$

EXPLORE 16A

▶ Substitution: One Variable Isolated

- When one variable is isolated, _____ is a common way to solve a system of equations.

Example 3

Solve the system of equations using substitution.

$a = \frac{1}{2}b - 1$
$4a + 3b = 15$

Plan Substitute the right side of the first equation into the second equation for a.
Solve for b.
Substitute the value for b in the first equation.
Solve for a.
Substitute the values for a and b into both equations to check.

Implement	Explain	Check
$4\left(\frac{1}{2}b - 1\right) + 3b = 15$	◀ Substitution	$4\left(\frac{9}{10}\right) + 3\left(\frac{19}{5}\right) = 15$
	◀ Distributive Property	
	◀ Addition Property of Equality	
	◀ Multiplication Property of Equality	
	◀ Substitution	
	◀ Least Common Denominator	
	◀ Alphabetically a is before b, so the ordered pair will be (a, b).	

☑ Checkpoint

Solve the system using substitution. Check your work.

$y = 10x - 6$
$5x - 2y = 7$

16A EXPLORE

▶ Substitution: No Variables Isolated

- If neither equation in a system has an isolated variable, a _____

  in either equation must be _____ in order to solve using substitution.

Example 4

Solve the system of equations using substitution.

$2x - 3y = 4$
$4x + y = -6$

Plan Isolate one of the variables.
Substitute the variable in the other equation.
Solve for the variable.
Then substitute that result into one of the equations.
Solve for the other variable.
Write the solution as an ordered pair.

Implement   **Explain**

$\qquad 4x + y = -6$

$\quad -4x \qquad -4x$ ◀ Addition Property of Equality

$\qquad\qquad y = -4x - 6$

$\qquad 2x - 3y = 4$ ◀ Substitution

◀ Distributive Property

◀ Addition Property of Equality

◀ Multiplication Property of Equality

◀ Substitution

◀ Addition Property of Equality

EXPLORE 16A

- Remember from Lesson 15, a system of equations can also have _____ (parallel lines) or _____ (coincident lines).

- When solving a system of _____ lines algebraically, the variables will simplify out, and the result will be an equation with no solution.

Example 5

Solve the systems of equations using substitution.

Parallel Lines:
$y = 3x - 1$
$y = 3x + 2$

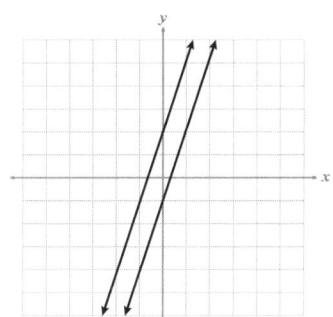

Coincident Lines:
$2x + y = 1$
$y = -2x + 1$

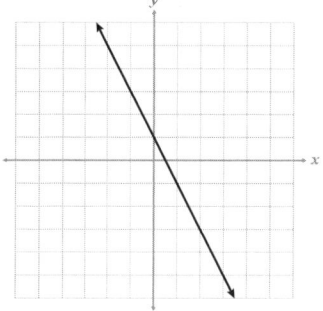

☑ Checkpoint

Solve the system using substitution.

$x - 2y = 7$
$x + y = 4$

Practice 1

Complete the problems on a separate sheet of paper.

Solve the system of equations using substitution.

1) ● + ● + ▲ + ▲ + ▲ = 25
   ● = − ▲ + 11

2) ● + ● + ● + ■ = −2
   ■ = − ● + 4

3) $y = 2x - 11$
   $y = \frac{1}{2}x + 4$

4) $a = 2b + 5$
   $a + b = 6$

5) $y = -3x - 2$
   $3x + y = -5$

6) $x + 2y = -4$
   $y = x + 1$

7) $y = \frac{1}{2}x$
   $x - 4y = -3$

8) $x - 7y = 0$
   $3x + 8y = -29$

9) $a - 8b = 1$
   $2a - 10b = 5$

10) $3x + y = 3$
    $7x + 2y = 1$

11) $4c - 2d = 0$
    $3c - 4d = -2$

12) $y = 4x - 4$
    $x = -\frac{1}{2}y - 2$

Mastery Check

Show What You Know

Tom is in Algebra class and writes an equation to represent his age relative to his brother's age. Tom's age is twice his brother's age plus one year. The equation below represents Tom (T) and his brother (b).

$$T = 2b + 1$$

A) The total age of Tom and his brother is 19.
Write an equation to represent Tom (T) and his brother (b).

B) Write a system of equations to represent the age of Tom and his brother. Remember to define the variables in the equations.

C) What are the ages of Tom and his brother? Solve the system of equations from part B.

D) Tom and his brother have an Uncle Kevin who is three times Tom's age minus Tom's brother's age. Find the age of Uncle Kevin. Show your work.

Say What You Know

In your own words, talk about what you have learned using the objectives for this part of the lesson and your work on this page.

16A PRACTICE 2

✏️ Practice 2

Complete the problems on a separate sheet of paper.

Solve the system of equations using substitution.

1) ■ + ■ + 3 = ▲
   ■ + ▲ + ▲ = 16

2) $x = \frac{1}{3}y + 2$
   $2x - 4y = 5$

3) $a = 4b - 10$
   $a = 3b + 11$

4) $y = -\frac{5}{2}x - \frac{5}{2}$
   $5x + 2y = -5$

5) $2x + 2y = 15$
   $y = 9x + 5$

6) $m = 6n$
   $2m + 8n = -15$

7) $4x + y = 5$
   $2x - 3y = 13$

8) $3e - f = -1$
   $e - 3f = 1$

9) $2p - q = -3$
   $8p - q = -7$

10) $50x + 25y = 200$
    $x + y = 1$

11) $e + 2f = 8$
    $2e - 3f = -19$

12) $5x - 4y = -1$
    $4y + 6 = 5x + 5$

Part B: Solving Systems Using Elimination

Objectives

In this part of the lesson, you will learn about solving systems using elimination.

By the end of this lesson, you will be able to do the following:

- Use elimination to solve a system of equations when one variable has a set of opposite coefficients.
- Use elimination to solve a system of equations by distributing -1 across an equation.
- Use elimination to solve a system of equations using linear combinations.

Why?

Using elimination to solve systems of equations is a very efficient method. It also helps reduce mistakes compared to other methods because you are eliminating a variable.

Warm Up

Simplify.

1) $3(2x - 4) + 8x$

2) $-15x + 4 + y - 2 + 15x$

3) $-(8x - 7) + 2(4x - 3) - 1$

▷ Elimination: Opposite Coefficients

- When a system of equations is written in _____,

  the _____ method is the most efficient way to solve.

- The Additive Inverse Property, _____, is used to eliminate the chosen variable.

- When the coefficients of a variable are opposites, they will simplify out of the system when _____ together.

16B EXPLORE

Example 1

Solve the system of equations using elimination.

$$10x - 14y = 28$$
$$-10x + 18y = 4$$

Plan Eliminate opposite x terms.
Solve for y.
Substitute y to solve for x.
Check solution.

Implement

$$\cancel{10x} - 14y = 28$$
$$+\ \underline{-\cancel{10x} + 18y = 4}$$
$$4y = 32$$
$$y = 8$$

Explain

◂ Elimination

◂ Multiplication Property of Equality

◂ Substitution

◂ Addition Property of Equality

◂ Multiplication Property of Equality

☑ Checkpoint

Solve the system of equations using elimination.

$$x + y = 13$$
$$x - y = 5$$

▶ Elimination: Multiplying by −1

- The _____ is key in solving systems using the _____ method.

- If the same number is multiplied across _____ of an equation, the equation maintains equality.

Example 2

Solve the system of equations using elimination.

$-3x + 8y = -10$

$2x + 8y = -15$

$\downarrow$

$-3x + 8y = -10$

$\underline{+2x + 8y = -15}$

$x + 16y = -25$ ✗

☑ Checkpoint

Solve the system of equations using elimination. Name the variable to be eliminated.

$2x - 3y = 6$
$7x - 3y = 16$

▶ Elimination: Linear Combinations

- When linear combinations are used, _____ equations in the system are _____ by a constant.

- Multiplying by a constant allows a variable to be _____ when the equations are _____ together.

16B EXPLORE

Steps to solving elimination:

1) Make sure equations are all written in the _____ form.

2) Choose the _____ to be eliminated and mark it (highlighter, arrow, etc.).

3) Determine the _____ of the coefficients for the variable to be eliminated.

4) Multiply one or more equations by the _____ to get the LCM. Make sure that the terms are opposites—one positive and one negative.

5) Add the equations together.

6) Solve for each variable one at a time.

7) Check by _____ the solution into both equations.

Example 3

Solve the system using linear combinations.

$8a + 3b = 4$
$5b = -34 + 7a$

$$8a + 3b = 4$$
$$-7a + 5b = -34$$

$$(-5)(8a + 3b = 4) = \quad -40a - 15b = -20$$
$$(3)(-7a + 5b = -34) = + \quad -21a + 15b = -102$$
$$\overline{}$$
$$-61a = -122$$
$$a = \frac{-122}{-61}$$
$$a = 2$$

☑ Checkpoint

Solve the system using linear combinations.
Name the variable to be eliminated and the least common multiple of the coefficients.

$6x + 2y = 33.50$
$5x + 7y = 45.25$

Practice 1

Solve using elimination. Name the variable you will eliminate before solving the problem.

1) $x + 2y = 2$
   $-x - 3y = 0$

2) $u + v = 66$
   $v = 30 + u$

3) $3a - b = 15$
   $-3a - 3b = -27$

4) $3x - 5y = 11$
   $3x + 3y = 3$

5) $5x + 12y = 19$
   $5x + 6y = 7$

6) ● + ● + ▲ + ▲ = 10
   ● − ▲ = 11

Solve each system using the elimination method. Name the variable you will eliminate and the least common multiple of the coefficients.

7) $x - y = 3$
   $4x - 3y = -6$

8) $x - 2y = 1$
   $2x + 3y = 16$

9) $5x + 6y = -8$
   $2x + 3y = -6$

10) $6x + 7y = 2$
    $5x + 8y = 6$

11) $4e - 1.5f = 6$
    $e + 3.5f = -14$

12) $7a + 14c = 2$
    $a = -\frac{12}{5}c$

Mastery Check

Show What You Know

A) Solve the system.
$4x - 2y = 55$
$x - y = 9$

B) Explain the steps you took to solve the system in part A.

C) Use the solution of the system in part A to determine the sum of x and y. Show your work.

D) Use the solution of the system in part A to determine the difference of twice y and x. Show your work.

Say What You Know

In your own words, talk about what you have learned using the objectives for this part of the lesson and your work on this page.

Practice 2

Complete the problems on a separate sheet of paper.

Solve each system using the elimination method.

1) $10x + 5y = -35$
$10x + 6y = -34$

2) $m - 2n = 25$
$m + 2n = 20$

3) $-0.6x + y = -3.5$
$x - y = 2.5$

4) $x + y = 4$
$x - y = 8$

5) $4x + y = 4$
$y = -12 + 2x$

6) ■ + ▲ = 18
■ + ■ + ▲ = 22

Solve using elimination. Name the variable you will eliminate and the least common multiple of the coefficients.

7) $2a - 15b = 7$
$a - 6b = 4$

8) $x + 3y = 6$
$3x + 4y = 13$

9) $3m = 5p + 8$
$3p = 2m + 3$

10) $10.5c + 8.5d = 74$
$c + d = 8$

11) $-4x + 2.5y = -21.75$
$2x + y = 7.5$

12) $16b + 3g = 30$
$22b + 5g = 43$

Targeted Review

In the Targeted Review, you will practice topics you have mastered in earlier lessons. Reviewing these concepts will help you be successful as you work through this unit.

Complete the problems on a separate sheet of paper.

1) Solve the system by graphing.

   $y = -\frac{1}{2}x + 5$

   $3x - 8y = 16$

2) Graph the system of inequalities.

   $y < -\frac{1}{2}x + 5$

   $3x - 8y < 16$

3) How do you determine the solution to a system of equations when graphing?

4) How do the graphs and solutions of systems of equations and systems of inequalities differ?

5) Solve by clearing the fractions: $\frac{2}{3}(5x + 8) = \frac{1}{15}$

6) Determine which point(s), in the set $\{(-6, 0), (8, -3), (16, -6), (-8, -9)\}$, are possible solutions to the linear equation: $3x - 8y = 48$

7) Zeke purchased veggies and dip each day at school as a snack for $1.75. After 11 days, Zeke had spent $19.25. Write the equation of the line. Choose the form that makes the most sense and name it.

8) The Juarez family used the equation $y = 0.32x + 85$ to represent the total cost of renting a car on their vacation. The family was charged for each mile they drove. Explain the meaning of the slope and the y-intercept in context.

9) Solve for the given variable w: $P = 2l + 2w$

10) Write the equation in standard form: $y - \frac{3}{5} = \frac{8}{3}(x + 9)$

Multiple Choice

 11) Determine the solution to the system given the graph.

   A) (0, 1)

   B) (0, −6)

   C) (0, 2)

   D) (2, 0)

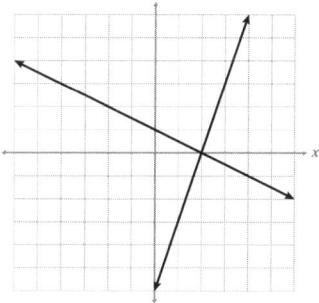

12) Select all possible solutions for systems of equations.

   ☐ one solution

   ☐ no solution

   ☐ infinite solutions

   ☐ all real numbers

Lesson 17
Applications of Linear Systems

Outline

Part A
Efficiently Solving Systems
- Efficiently Solving Systems
- Extending Solutions

Part B
Applications of Linear Systems
- How Much or How Many
- Coins
- Wind and Water

Targeted Review

Vocabulary

There are no new vocabulary words for this lesson.

17A EXPLORE

Part A: Efficiently Solving Systems

Objectives

In this part of the lesson, you will learn about efficiently solving systems.

By the end of this lesson, you will be able to do the following:

- ✓ Choose the best method to solve a system of linear equations and justify your choice.

Why?

What is the most efficient way to solve a system of linear equations? Can you find the answer in fewer steps? Deciding the most efficient method to solve will help you become a stronger math student for this and future math courses.

Warm Up

1) What are the three methods you have learned to solve a system?

2) Name the three types of solutions to a system. Make a sketch of each.

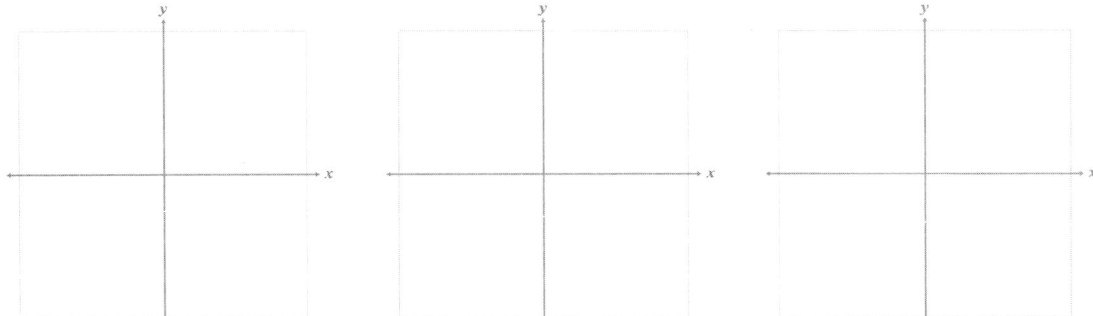

▶ Efficiently Solving Systems

- If a problem does not specify which method to use to solve the system of equations, you must determine which method is most _____ by _____ the equations.

EXPLORE 17A

Example 1

Solve the system of equations using graphing and one other method. Explain which method is more efficient and why.

$y = 2x + 3$
$y = 5x + 7$

Method:

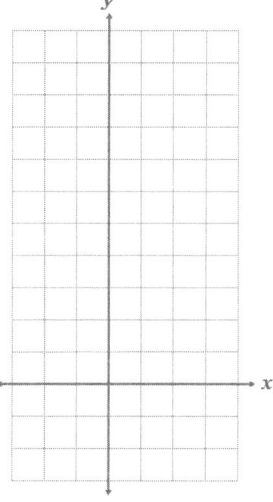

Explain

- In this case, substitution was more _____ than graphing because an _____ could not be found by graphing alone.

- Substitution resulted in a _____ solution since the solution contained fractions.

- Substitution was more efficient than elimination because a _____ was already _____ and could be quickly substituted into the other equation.

☑ Checkpoint

Given the following system, explain what method would be most efficient to use and why. (Do not solve the system.)

$2x - 7y = 14$
$3x + 5y = 15$

17A EXPLORE

▶ Extending Solutions

- Sometimes, you will be asked a question that cannot be answered until _____ _____.

- First, solve the _____ using any method. Then, use the _____ found to complete the problem.

Example 2

Determine the quotient of y and x for the system of equations.

$y = 6x - 5$
$2x - 4y = -13$

Plan Solve for x and y using substitution.
Calculate the quotient.

Implement

☑ Checkpoint

A) Name the solution.

B) Determine the difference between y and x.

C) Determine the quotient of y and x.

D) Determine $3x + y$.

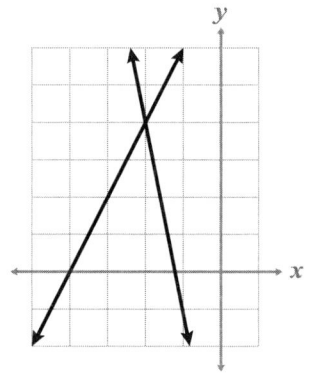

Practice 1

Complete the problems on a separate sheet of paper.

1) Graph.
   $y = \frac{1}{2}x - 2$
   $y = 2x - 15$

2) Solve the system from problem 1 using a different method.

3) Look at problems 1 and 2. Which method of solving (graphing or your chosen method) was more efficient for the system? Explain.

Use the most appropriate method to solve the system of equations.

4) $2x + 3y = 0$
   $-5x + 3y = -21$

5) $y = 2x + 8$
   $y - 4 = x + 2$

6) Compare the method(s) used to solve problems 4 and 5. What method did you use to solve each problem? Why did you choose that method?

---

7) Given the graph of the system, name the solution. Then, determine $2x - y$.

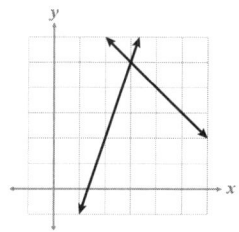

Match the system to the correct solution.

_____ 8) $x + 3y = 37.5$
          $4y - x = 39.5$

_____ 9) $2x + 3y = 6$
          $x = -\frac{3}{4}y$

A) $(-3, 4)$
B) $(11, 4.5)$
C) $(4, -3)$
D) $(4.5, 11)$

10) Use the solution to problem 8 to determine: $x - y$.

11) Use the solution to problem 9 to determine: $|x| - 3y$.

Explain when it is most efficient to use each method of solving a system.

12) Graphing

13) Substitution

14) Elimination

17A MASTERY CHECK

Mastery Check

Show What You Know

A city was overlaid onto the coordinate plane. Traffic lights are being installed at the intersection of Pine Street and Mulberry Street.

The equations of two intersecting streets are given below.

Pine Street: $2x + 3y = 12$

Mulberry Street: $-4x + 3y = -12$

A) Graph the equations using the x- and y-intercepts to show the streets and the intersection.

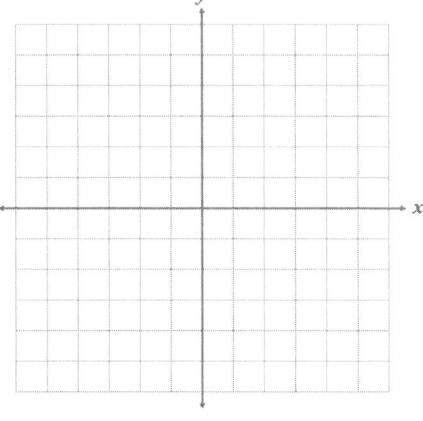

B) The city planner wants the exact location where the streets intersect. Explain why this graph will not give the exact location.

C) Solve the system using another method to find where the traffic lights will be located.

Say What You Know

In your own words, talk about what you have learned using the objectives for this part of the lesson and your work on this page.

Practice 2

Complete the problems on a separate sheet of paper.

1) Explain why graphing *would not* be the best approach to solving the system.

   $y = 10x + \frac{8}{5}$
   $y = -5x - \frac{7}{5}$

Solve using any method. State your chosen method before solving.

2) $y = 10x + \frac{8}{5}$
   $y = -5x - \frac{7}{5}$

3) $2x + y = 8$
   $y = -2x - 11$

4) How can you show what problem 3 *looks like* to prove that your solution is correct?

5) Solve.
   $c - d = 0$
   $c + 4d = 200$

6) What method did you use to solve problem 5? Why did you choose that method?

Match the most efficient method of solving to the system.
Answer choices can be used more than once.

_____ 7) $5x - y = 58$
        $8x - y = 97$

_____ 8) $y = -2x - 3$
        $4x + y = -5$

_____ 9) $4x - 2y = 55$
        $x - y = 9$

A) Graphing
B) Substitution
C) Elimination

10) Solve.
    $5x - y = 58$
    $8x - y = 97$

11) Determine the difference between y and x from problem 10.

12) Solve.
    $y = -2x - 3$
    $4x + y = -5$

17B EXPLORE

Part B: Applications of Linear Systems

Objectives

In this part of the lesson, you will learn about applications of linear systems.

By the end of this lesson, you will be able to do the following:

- ✓ Write systems of equations for how much and how many, coin, and wind and water word problems and solve them.
- ✓ Explain the solutions to how much and how many, coin, and wind and water problems and determine if the solutions are reasonable.

Why?

You have a bill and know the number of items you purchased but what was the individual cost per item? Being able to apply linear systems to real-world problems will help you solve situations like this one and many more.

Warm Up

Translate the words into an algebraic expression or equation.

1) Nine times some number less two.

2) Two times the sum of x and y is eleven.

3) The quotient of a number and three is seven

How Much or How Many

- Which type of word problem gets its name from the question being asked?

  _____

- When solving these types of problems, remember the following:

  - You need at least two _____ to solve a system.

  - Both _____ need to be defined.

  - Any _____ can be used to solve for the correct solution.

EXPLORE 17B

Example 1

Use the given information to define your variables, write a system of equations, and solve.

A math test is worth a total of 100 points and contains 24 questions. The test includes some worked solution questions worth 5 points each and some multiple-choice questions worth 4 points each. How many of each question type are on the test?

Plan Define your variables. m:
Write the equations. w:
Solve.

Implement

$ = $ 100 points

$ = $ 24 questions

Explain

> Remember to write your answer in a short sentence. Try saying the answer out loud to help you determine if the answer is possible and to help you find any errors.

Example 2

Use the given information to define your variables, write a system of equations, and solve.

The Humanitarian Club is selling two different types of bracelets for a charity. A bracelet with 4 red beads and 12 blue beads costs $40. A bracelet with 8 red beads and 8 blue beads costs $32. How much does each color of bead cost?

Plan Define your variables. **Implement**
Write the equations. r: b:
Solve.
$ = $ $40 bracelets
$ = $ $32 bracelets

Explain

17B EXPLORE

> ☑ **Checkpoint**
>
> A car wash was held for a local charity. A total of 38 cars (c) and trucks (t) were washed, and $336 was raised. If it cost $8 to wash a car and $12 to wash a truck, how many cars were washed? How many trucks were washed?

▶ Coins

- Word problems that mention _____ are a special type of "how much/how many" problem.

- For coin problems, you must give extra attention to _____.

- When solving coin problems, write all monetary values in terms of _____.
    - Nickel: _____
    - Dime: _____
    - Quarter: _____

Example 3

Define your variables, write a system of equations, and solve.

A coin jar contained $6.30 in nickels and quarters. If there were 26 coins, how many quarters were in the jar?

n: _____ q: _____

This problem has two total values: _____ and _____.

$$(-0.05)(n + q = 26) = -0.05n - 0.05q = -1.30$$
$$ + 0.05n + 0.25q = \underline{6.30}$$

> You could also clear the decimals before solving by multiplying by the LCD, in this case 100.

EXPLORE 17B

☑ Checkpoint

A coin jar contained $1.20 in nickels and dimes.
If there were 18 coins, how many nickels and dimes were in the jar?

▶ Wind and Water

- Wind and water problems usually involve solving for the _____, _____,

  or _____ of a plane in the wind or a boat in the water.

- The formula for wind and water problems is _____.

- The rate, r, is the _____ of the speed of the plane and the

  wind or the boat and the water.

> You may often see the formula $d = rt$. The order of rate and time does not affect the final values, but placing time first for this type of problem can make it simpler to distribute the values.

- Wind and water words used to describe rates (r):

r: rate	airplane	boat/kayak/canoe	effect
	headwind	upstream (against the current)	slows down (plane/boat)
	tailwind	downstream (with the current)	speeds up (plane/boat)

- Time increments will be in terms of _____ since the speed of something is

  usually miles per hour (mph) or kilometers per hour (km/h).

17B EXPLORE

Example 4

Write a system of equations and use it to answer the given questions.

When flying into a headwind, a plane took 75 minutes to travel 450 km. With a tailwind, the same trip was 15 minutes shorter. What was the speed of the plane and the airspeed?

Define variables:
p: plane, w: wind

Convert all time to hours:
$75 \text{ min} = 1\frac{1}{4} \text{ hr} = \frac{5}{4} \text{ hr}$
$75 - 15 = 60 \text{ min} = 1 \text{ hr}$

A) Create the system of equations using known facts:

d	t	rate
450	$\frac{5}{4}$	$(p-w)$ headwind
450	1	$(p+w)$ tailwind

$450 = \frac{5}{4}(p-w)$ The equation for the trip in a headwind.
$450 = 1 \cdot (p+w)$ The equation for the trip in a tailwind.

Note that the distance is the same in both equations because the only difference between trips was the rate and the time.

$\left(\frac{4}{5}\right) 450 = \left(\frac{4}{5}\right)\left(\frac{5}{4}(p-w)\right)$

$360 = p - w$
$+\ 450 = p + w$
$810 = 2p$
$p = 405$

$450 = p + w$
$450 = (405) + w$
$w = 45$

The plane's speed was 405 km/h, and the wind speed was 45 km/h.

B) What is the rate when the plane has a tailwind? $r = p + w$ in a tailwind

C) What is the rate when the plane encounters a headwind? $r = p - w$ in a headwind

D) What is the speed of the plane in still air (no wind)?

EXPLORE 17B

Example 5

Write a system of equations and use it to answer the given questions.

The Chu family went on a canoe tour of Beautiful Canyon. The twelve-mile trip took four hours to travel upstream and 1 hour and 20 minutes to travel downstream. How fast was the family paddling? How fast was the water on the day of their trip?

p: canoe, w: water

1 hour 20 minutes =

$d = tr$

d	t	rate
		heading upstream
		heading downstream

☑ Checkpoint

A plane flew 450 km/h. The wind speed was 25 km/h.

A) What is the speed of the plane in still air?

B) How fast would the plane be traveling with a tailwind?

C) How fast would the plane be traveling with a headwind?

D) What is the flight time with a tail wind for a distance of 1,900 km?

E) What is the flight time for a plane with a headwind traveling 850 km?

17B PRACTICE 1

✏️ Practice 1

Complete the problems on a separate sheet of paper.

Write the system of equations, define your variables, and solve.

1) Bailey was at the farmers' market. She wanted to buy tomatoes and garlic to make pasta sauce. At the first farm stand, she purchased 8.5 pounds of tomatoes and 0.5 pounds of garlic for $26.75. At the second farm stand, she purchased 12 pounds of tomatoes for $30. What was the price per pound for tomatoes? What was the price per pound for garlic? (Assume the price per pound of tomatoes was the same at both stands.)

2) The sum of the ages of Balthazar's two grandmothers is 167. The difference between their ages is 7. How old are Balthazar's grandmothers?

3) Xavier purchased two pairs of jeans and five shirts for $98. Yoel purchased three pairs of jeans and seven shirts for $141. What would the cost be for one pair of jeans and one shirt?

4) The music hall sells two ticket levels for concerts, floor and mezzanine. A total of 525 seats were sold for the Saturday night show. There were twice as many mezzanine seats as floor seats. How many floor tickets were sold? How many mezzanine tickets were sold?

5) When the birth years of Mark and Steven are added, the sum is 3,974. Steven was born 6 years after Mark. What year was each person born?

6) A math test worth 100 points and containing 30 questions was assigned to the class. Some questions were worth 2 points, and some were worth 10 points. How many questions of each type were on the test?

Malia rows her kayak at a rate of 3 km/h, and the current is 2 km/h.

7) How fast would Malia be able to move upstream?

8) How fast would Malia be able to move downstream?

9) How long will it take to travel 5 kilometers upstream and back? Explain.

A crew team was rowing 6 mph in a waterway with a current of 3 mph.

10) How fast would the team be able to move upstream?

11) How fast would the team be able to move downstream?

12) How long will it take to travel 18 miles upstream and back? Explain.

Write the system of equations, define the variables, and solve.

13) Josef collected dimes and nickels. He had 30 coins that totaled $2.10. How many of each coin does Josef have?

14) When Calia went to the post office, the price of a stamp for a letter was $0.55, and the price of a stamp for a postcard was $0.35. If Calia bought twenty fewer postcard stamps than letter stamps and spent $38 total, how many of each type of stamp did Calia buy?

15) Logan has a total of 20 bills in his wallet. He has ten-dollar bills and twenty-dollar bills totaling $330. How many of each bill does Logan have?

16) An airplane traveling 2,400 miles from Philadelphia to Los Angeles takes 6 hours. The return trip takes 5.5 hours. What is the speed of the plane? What is the average speed of the wind? Round to the nearest unit.

Mastery Check

Show What You Know

The combined age of Sara and Elizabeth was 72. If Elizabeth's age were doubled, she would be 30 years older than Sara.

A) Define your variables and write a system of equations to represent the given scenario.

B) What are the ages of each person? Use your system from part A to solve.

C) Sara and Elizabeth emptied their pockets of all their loose change and found 13 coins. They only had dimes and quarters totaling $2.65. Define your variables and write a system of equations representing the coins Sara and Elizabeth have. Do not solve.

Say What You Know

In your own words, talk about what you have learned using the objectives for this part of the lesson and your work on this page.

17B PRACTICE 2

✏️ Practice 2

Complete the problems on a separate sheet of paper.

Write the system of equations, define your variables, and solve.

1) A math test has sixteen questions made of multiple-choice questions and open-response questions. There were three times as many multiple-choice questions as open-response questions. How many multiple-choice questions and open-response questions were each on the test?

2) The difference between the ages of Tori and Thomas is three years. One-third of Tori's age minus Thomas' age is equal to negative five years. How old are Tori and Thomas?

3) The Simensons went to the store to purchase food for school lunches. On their first trip, they purchased 5 pounds of apples and 6 bags of spinach for a total of $37.95. For their next trip, they purchased 5 bags of spinach and 3 pounds of apples for $28.72. Assuming the prices were the same for each trip, what is the cost of a pound of apples? What is the cost of a bag of spinach?

4) An artist needed more supplies to finish a project. They purchased two brushes and five tubes of paint for $52.25. Then, they purchased one brush and seven tubes of paint for $58.75. What was the cost of a tube of paint for the project?

5) Magda has a younger brother and younger sister. The sum of their ages is four. Her brother is twice her sister's age minus two. How old are Magda's siblings?

6) The Kelly family and the Miller family went to brunch. Adult meals cost $25, and meals for children under 12 were $16. The two families spent a total of $219. If six more children than adults attended, how many meals were purchased for children?

7) Ella and Claire had a combination of 18 dimes and nickels totaling $1.25. How many of each coin did the girls have?

A plane flew 575 mph. The wind speed was 35 mph.

8) How fast would the plane be traveling with a tailwind?
9) How fast would the plane be traveling with a headwind?
10) What is the flight time with a tail wind for a distance of 1,525 miles?
11) What is the flight time for a plane with a headwind traveling 1,755 miles?

Write the system of equations, define the variables, then solve.

12) A park ranger traveled across a lake and back in his boat. He traveled 32 kilometers in one direction, which took 4 hours. His return trip was the same distance and time. What was the speed of his boat? What was the speed of the water?

13) A plane trip from New York to London takes 7 hours. The return flight takes approximately 8 hours to fly the same distance of 3,500 miles. What is the speed of the plane?

14) Fabian was counting his commission for doing chores for the past three months. He was paid in five-dollar bills and one-dollar bills. Fabian earned $83. The number of five-dollar bills was six times the number of one-dollar bills, less two. How many of each bill does Fabian have?

15) Burt has $4.40 in dimes and half-dollar coins. The difference in value between half-dollars and dimes is $2.60. How many of each coin does Burt have?

TARGETED REVIEW 17

◎ Targeted Review

In the Targeted Review, you will practice topics you have mastered in earlier lessons. Reviewing these concepts will help you be successful as you work through this unit.

Complete the problems on a separate sheet of paper.

1) Graph the system of inequalities.
$y > 2x - 3$
$y \leq x + 4$

2) Name the quadrant or quadrants where the solution to the system in problem 1 is located.

3) Solve the system of equations.
$x + y = 3$
$x = -\frac{1}{3}y - 2$

4) Solve the system of equations.
$7x + 2y = 12$
$7x + 4y = 3$

Classify the numbers by their most specific name.

5) $(-7)(-2.3)$:

6) $\frac{1}{2}\sqrt{2}$:

7) $83 - 83$:

8) $-26 + 3$:

natural	whole
integer	rational
irrational	real

Classify the equations of lines given in the box as parallel, perpendicular, or neither.

9) $f(x)$ and $g(x)$

10) $g(x)$ and $j(x)$

11) $h(x)$ and $j(x)$

12) $f(x)$ and $h(x)$

13) $g(x)$ and $h(x)$

$f(x) = \frac{3}{2}x + 7$
$g(x) = -\frac{3}{2}x - 3$
$h(x) = \frac{3}{2}x + 1$
$j(x) = \frac{2}{3}x - 3$

Convert minutes into hours or fractions of an hour.
Write fractions greater than one as an improper fraction.

14) 45 minutes:

15) 100 minutes:

16) 42 minutes:

17) 75 minutes:

TARGETED REVIEW 17

18) Solve. Justify your steps with the properties.

$4(9 - x) + 6x = 20$

19) Saul was going on a road trip. He started his trip at his house and drove 55 miles per hour. He used the equation $d(t) = 55t$ to find how many hours, t, it would take to travel various distances, $d(t)$. Explain the meaning of the slope and the y-intercept from the word problem.

20) Find the x- and y-intercepts for the equation: $3x - 15y = 45$

Multiple choice

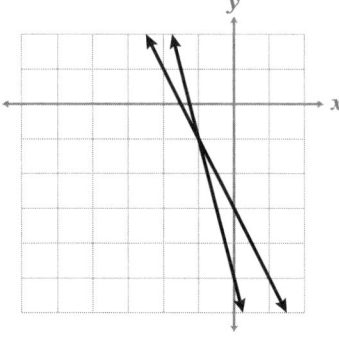

_____ **21)** Name the solution to the system of equations.

    A) $(-1, 1)$

    B) $(-1, -1)$

    C) $(1, -1)$

    D) $(1, 1)$

_____ **22)** Find the product of x and y using the solution to the system of linear equations.

$y = 2x - 3$
$y = x + 4$

    A) -12

    B) -4

    C) 18

    D) 77

Lesson 18
More Applications of Linear Systems

Outline

Part A
More Applications of Linear Systems

- Geometry Formulas
- Break-Even
- Digit

Part B
Applications of Linear Inequalities

- Writing a Linear Inequality in One Variable
- Writing Systems of Inequalities

Targeted Review

Vocabulary

There are no new vocabulary words for this lesson.

18A EXPLORE

Part A: More Applications of Linear Systems

Objectives

In this part of the lesson, you will learn about more applications of linear systems.

By the end of this lesson, you will be able to do the following:

- ⊘ Write systems of equations for break-even, formula, and digit word problems and solve them.
- ⊘ Explain the solutions to break-even, formula, and digit problems and determine if the solutions are reasonable.

Why?

Suppose you invested a certain amount of money into starting a business. How long would it take for you to earn the money you invested back and make a profit? You can answer this type of question and more by using systems of equations.

☁ Warm Up

1) How can you find the perimeter of a figure?

2) What is the formula for the perimeter of a rectangle?

3) An isosceles triangle has two sides of equal length and a third side of a different length. Write a formula for the perimeter of an isosceles triangle.

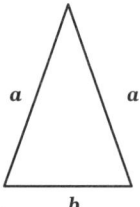

▶ Geometry Formulas

■ Perimeter measures the total distance _____ the edge of an object.

EXPLORE 18A

Example 1

Write a system of equations. Solve.

When two sides of a triangle are the same length you have an isosceles triangle. An isosceles triangle has a perimeter of 18 centimeters. The sum of the two equal sides is 2 more than the remaining side. What are the side lengths of the triangle?

Plan Sketch a triangle.
Define the variables for side length.
Write a system of equations.
Find the side lengths.

Implement

x: side of triangle (2), y: side of triangle (1), $P = 18$ cm

$x + x + y = 18 \to 2x + y = 18$

$x + x = y + 2 \to 2x = y + 2$

Explain

Example 2

Write a system of equations. Solve.

The perimeter of a rectangle is 26 centimeters. Twice the width of a rectangle minus the length is -1. What are the dimensions of the rectangle?

Plan Define the variables.
Substitute the known facts into the perimeter formula of a rectangle.
Complete the system of equations.
Solve.

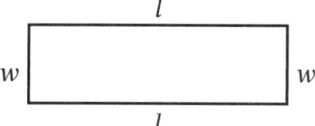

Implement

18A EXPLORE

> ### ☑ Checkpoint
>
> **Write the system of equations. Do not solve.**
>
> The perimeter of a rectangle is 48 centimeters. The width is half of the length plus 3 centimeters.

▶ Break-Even Problems

- To _____ in business means that you have made back all of the money, or expenses, that were spent to start the business or keep it running.

- Break-even problems can also be used to _____ different choices to see what the best deal or outcome is.

Example 3

Write a system of equations. Solve.

Gregg needed to know the break-even point for his sticker business. His expenses totaled $5,000 plus $1 per sticker made. He planned to sell each sticker for $7. How many stickers will Gregg need to sell to break even? (Only whole numbers of stickers can be sold.)

Plan

Define your variables.
Write a system of equations.
Solve.

Implement

x: stickers, y: costs
expenses: $y = 5{,}000 + 1x$
income: $y = 7x$

Example 4

Write a system of equations. Solve.

A family is comparing the cost of a graduation party at two different locations. The fire hall charges $5 per person and a room fee of $350. The restaurant charges $10 per person and a room fee of $225.

A) How many people would need to attend for the cost of the two locations to be equal?

x: people, y: cost

Fire hall: $y = 5x + 350$

Restaurant: $y = 10x + 225$

B) When is the fire hall a better deal for the family? When is the restaurant a better deal? Provide examples to back up your reasoning.

- The fire hall is a better deal when _____ 25 people attend the party.

- The restaurant is a better deal when _____ 25 people attend.

> Break-even problems are usually written (x, y) because a graph can be made to show the break-even point if necessary.

☑ Checkpoint

Write a system of equations representing the expenses and income for the given scenario. Do not solve.

Pat's Provisions has monthly expenses including $3 per sandwich sold and $1,200 in additional expenses. The store sells each sandwich for $8.50.

(x: sandwich, y: cost)
expenses: _____   income: _____

18A EXPLORE

▶ Digit Problems

- Digit problems require that special attention is paid to the _____ of a number.

- The problems in this section will use _____, t, and _____, u.

> Make sure to carefully write the variable t. It is not commonly used, but for digit problems, it makes the most sense.

Example 5

Write a system of equations. Solve.

The sum of the digits in a two-digit number is 10. When the value of the digits are reversed, the number is 72 more than the original number. What is the number?

Implement

t: tens, u: units

$$t + u = 10$$
$$t + 10u = 10t + u + 72$$
$$-9t + 9u = 72$$
$$\left(\tfrac{1}{9}\right)(-9t + 9u = 72) = -t + u = 8$$
$$+t + u = 10$$
$$2u = 18$$
$$u = 9, t = 1$$

Explain

Using mental math and replacing u with 9 in the original equation $t + u = 10$, you find that $t = 1$.

When the digits are placed in the correct place value location: _____

Check that the digits are written in the correct order.

Remember, the two numbers that are being solved for create a single two-digit number.

☑ Checkpoint

Write a system of equations. Do not solve.

When the tens digit is subtracted from the units digit the result is 5. The number with the digits reversed is one more than ten times the units.

Practice 1

Complete the problems on a separate sheet of paper.

Write a system of equations, define your variables, and solve.

1) Two angles that are supplementary total 180 degrees. The smaller angle is half the larger angle less 18. What are the measures of each angle?

2) The perimeter of a rectangle is 48 centimeters. The width is one-third the length. What are the dimensions of the rectangle?

3) An isosceles triangle has a perimeter of 41 yards. Twice the sum of the sides with different lengths is 52. What is the length of each side of the triangle?

4) Two angles that are complementary total 90 degrees. Twice the smaller angle plus three times the larger angle is 240. What is the measure of each angle?

5) The perimeter of a rectangle is 22 inches. The difference between the length and the width is five. What are the dimensions of the rectangle?

6) What would the area be of the rectangle in Problem 5? Write the formula and solve.

7) The sum of two numbers is −7. The second number is five times the first number less 13. What are the two numbers?

8) The sum of the digits of Joan's age is 16. The sum of the tens digit and twice the units digit is 25. How old is Joan?

9) The units digit of a number subtracted from the tens digit is −3. The value of the digits reversed is twenty times the tens digit plus three. What is the number?

10) Tanya's T-shirts purchased blank t-shirts for $3.25 each and screen printing materials for $180. Tanya sells her designed t-shirts for $12 each. How many shirts must Tanya's T-shirts sell to break even?

11) Bob's Biscuits spent $500 on baking supplies. Each biscuit sold at $2.50 each. How many biscuits will Bob need to sell to break even?

12) If Bob increased the selling price to $3 per biscuit, how many fewer biscuits would he need to sell to break even?

18A MASTERY CHECK

Mastery Check

Show What You Know

Quinn started a healthy dinner club meal service. The outgoing expenses were $300 for supplies and $8 per meal. Quinn sold his kits for $25 per kit.

A) Write a system of equations for expenses and income. Define your variables.

m: meal, c: costs

expenses:

income:

B) Solve the system from part A for the number of meal kits Quinn needs to sell to break even. Explain your steps.

C) How much will Quinn earn if he sells 42 meal kits? Show your work. (Hint: earnings represent the money left after expenses are deducted.)

Say What You Know

In your own words, talk about what you have learned using the objectives for this part of the lesson and your work on this page.

PRACTICE 2   18A

✏️ Practice 2

Complete the problems on a separate sheet of paper.

Write a system of equations, define your variables, and solve.

1) Two sides of a triangle are of equal length. The remaining side is one-fourth as long as one of the equal sides, less two. If the perimeter of the triangle is 34 inches, what is the length of each side?

2) Two complementary angles total 90 degrees. The larger angle is three times the smaller angle, plus two. What is the measure of the larger angle?

3) Two supplementary angles total 180 degrees. When the smaller is subtracted from one-fourth the larger angle, the result is −55. What is the difference between the two angles?

4) The length of a rectangle is five times the width. If the perimeter is 72 feet, what are the dimensions of the rectangle?

5) Another rectangle has a perimeter of 72 feet. The length is six more than the width. What are the dimensions of this rectangle?

6) Find the positive difference between the areas of the rectangles in problems 4 and 5.

7) The sum of the digits of a two-digit number is 11. When the digits are reversed, the sum is 27 more than the original number. What is the number?

8) Gerald is thinking of two numbers. The difference between the numbers is 25. The sum of the numbers is 61. What are the two numbers?

9) The Corvea family saved $1,200 for their yearly vacation. The hotel costs $125 per night and they have $500 in travel costs. Write a system to determine the number of nights the family can stay at the hotel. Will the family have any money left over?

10) The Corvea family found another hotel that costs $80 per night but their travel expenses would be $650. Would the family be able to stay longer even though the travel expenses increased?

11) Chrissy was overcharged 18 cents when the cashier reversed the digits for her tax owed. If the sum of the digits is 10, what is the correct tax?

12) Cami's Cleaners charges $40 for every office they clean. This month they purchased $2 of supplies per office and spent an additional $225 on travel costs. How many offices will Cami's Cleaners need to clean to break even this month?

18B EXPLORE

Part B: Applications of Linear Inequalities

Objectives

In this part of the lesson, you will learn about applications of linear inequalities.

By the end of this lesson, you will be able to do the following:

✓ Identify and apply the following key words to write an inequality from a verbal model:
- is more than
- is less than
- is no more than
- is at least
- is at most
- has a minimum of
- has a maximum of

✓ Write a system of inequalities from a scenario and solve for the solution.

✓ Explain what the solution to a system of inequalities means within a given context.

Why?

Your family is planning a vacation. They save a certain amount each month but have a maximum spending limit for the trip. How many days can the trip be? Do you have to spend all of the money saved? Solving systems of linear inequalities will help you answer questions like these.

Warm Up

Determine which ordered pairs, if any, are solutions to the system of inequalities. Write "yes" or "no" for each pair.

$x + y < 7$

$x - 5 > -5$

1) $(0, 5)$ 2) $(-2, 5)$ 3) $(2.33, 6.78)$

▶ Writing a Linear Inequality in One Variable

Symbol	Common Wording	Additional Wording
>	is greater than	
≥	is greater than or equal to	
<	is less than	
≤	is less than or equal to	

Example 1

Highlight the phrase that represents the inequality symbol needed. Write the inequality symbolically. Do not solve.

A) Ethan needs to work ==at least== 30 hours this week. $h \geq 30$

B) Daryl can reformat no more than 5 computers a week.

C) Kricket saves $15 per week for family birthdays throughout the year. She needs a minimum of $500 this year.

Example 2

Write a compound inequality and solve.

Rabina wants to determine the number of hours she would need to work to make a minimum of $1,200. She makes $16 per hour. Her boss told her that she could make no more than $2,560 in a given month. What are the possible hours that Rabina can work?

Plan   Define your variables. h: hours
Identify key words.
Write the system of inequalities.
Solve.

Implement

$16h \geq 1{,}200$ AND $16h \leq 2{,}560$ ◀ Multiplication Property of Equality

Explain

☑ Checkpoint

Write an inequality. Do not solve.

A) Pippa wants to earn a minimum of $1,200 this month. She makes $15 per hour.

B) Pippa would prefer to work no more than 70 hours a month.

▶ Writing Systems of Inequalities

- When writing a system of inequalities, you will often need _____ inequalities to represent a real life situation.

18B EXPLORE

Example 3

Write a system of inequalities. Shade only the solution on the graph.

The Nor family wants to include fruits and vegetables with each lunch this week. Fruit salad, x, is $6 per pound and carrots, y, are $2 per pound. They can spend no more than $18 this week on these items.

A) x: fruit salad, y: carrots   fruit: carrots:

   spending limit: $6x + 2y \leq 18$

B) Solve for the x-intercept   Solve for the y-intercept
    when y is zero.   when x is zero.

    $(a, 0)$ $(0, b)$

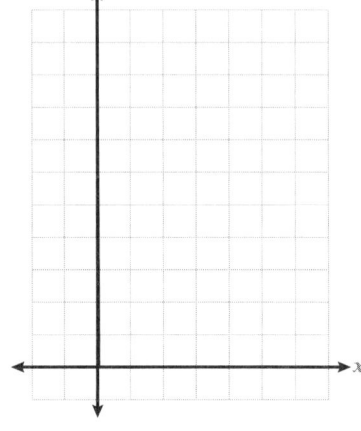

C) What does the shaded region represent?

    The shaded region represents the possible

    _____ of fruit salad

    and carrots that the Nor family can purchase for _____ $18.

D) Would it be possible for the Nors to purchase 3 pounds of fruit salad and 2 pounds of carrots?

E) Would the family be able to purchase a fraction of a pound of fruit salad and/or carrots?

☑ Checkpoint

The vertices of a rectangle are $(1, 6), (-2, 6), (-2, -3),$ and $(1, -3)$.

Graph the vertices and write a system of inequalities that will shade the area of the rectangle.

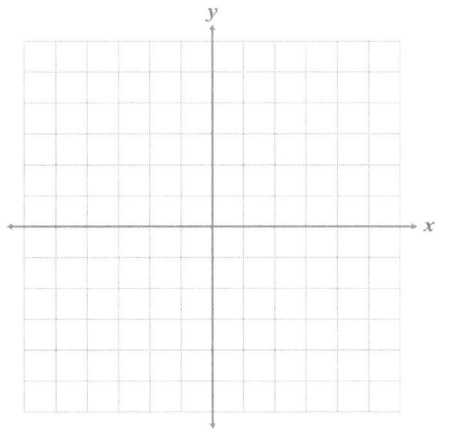

Practice 1

Complete the problems on a separate sheet of paper.

Identify the phrase that determines the inequality symbol and highlight or underline it. Write a single-variable or compound inequality and solve. Write a sentence to explain your solution within the given context.

1) Khaled could spend no more than $100 a month on gas. If gas costs $2.75 per gallon, how many gallons of gas can Khaled purchase?

2) Students must have indoor recess at school when the temperature is below 32° or above 93°. Write a compound inequality to show when students cannot have recess outside. When can students have outdoor recess?

3) Giorgi earns a minimum of $300 per week being paid $25 per hour. Giorgi can earn no more than $1,000 each week due to work guidelines. What is the range of time that Giorgi can work in one week?

Define your variables for the system. Write a system of inequalities. Do not solve.

4) The Adams family read a chapter of a book together each night before bed. Depending on the number of pages in the chapter, the family would spend between 15 minutes and 45 minutes reading. The chapters of the books are all at least 2 pages. No chapter is longer than 23 pages.

5) The school was hosting a field trip. At least 4 students must attend the field trip. At most 48 students could attend. At least one chaperone must attend for any student trip to occur.

6) Reema works part-time at a local coffee shop. Reema earns $9 per hour and can work a maximum of 20 hours each week. Reema must work at least 4 hours each week.

7) Write the system of inequalities. A rectangle has vertices at $(4, -2)$, $(4, 3)$, $(-1, 3)$, and $(-1, -2)$. Graph the vertices, then write a system of inequalities that represents the sides of the rectangle and shades the area of the rectangle.

8) What does the shaded region represent for the rectangle in problem 7?

9) Melody practices piano and vocals for a minimum of 12 hours each week. Her combined practice time for both cannot exceed 15 hours a week because of school work. Melody needs to practice piano at least zero hours a week. She also needs to practice vocals at least zero hours each week. Write a system of inequalities to represent all possible practice times where piano is the x-axis and vocals is the y-axis.

10) Graph the solution to problem 9. What does the shaded region represent for Melody?

11) Hermine works two part-time jobs while in college. Her catering job (x) pays $8 per hour, and her tutoring job (y) pays $20 per hour. Hermine must earn at least $250 total from both jobs each week, and she can work at most 15 hours per week. Write a system of inequalities to represent all possible ways for her to work both jobs and earn at least $250.

12) Graph the system of inequalities from problem 11. What is the fewest number of tutoring hours Hermine can work? What are two possible combinations of hours that Hermine can work? (Round to the nearest whole hour.)

18B MASTERY CHECK

Mastery Check

Show What You Know

A small music shop specializes in creating two Chinese instruments called the xiao and yaogu. At most 9 xiao can be created each month at the music shop. The music shop makes no more than 12 yaogu each month.

A) Write a system of inequalities that represents the number of each instrument that can be made each month. Remember to include the minimum number that can be made each month as well.

Each month, a group of musicians comes to the music shop to test the new instruments. The musicians are allotted 2 hours of playing time for each xiao and 3 hours of playing time for each yaogu. The number of combined hours the group spends testing all the instruments cannot exceed 42.

B) Write an inequality to represent all possible times it takes musicians to *test* new instruments.

C) Graph all inequalities in parts A and B on the same coordinate plane. Mark all intersection points that enclose the shaded region.

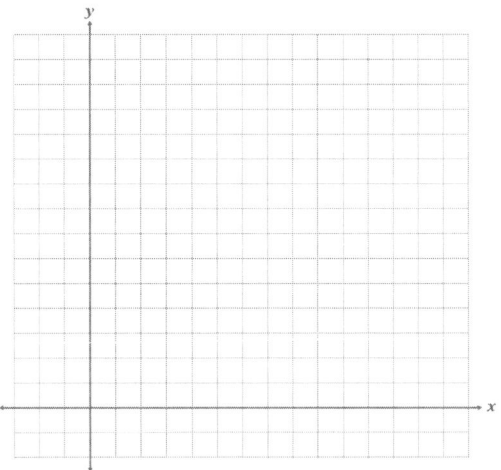

Say What You Know

In your own words, talk about what you have learned using the objectives for this part of the lesson and your work on this page.

Practice 2

Complete the problems on a separate sheet of paper.

Identify the phrase that determines the inequality symbol and highlight or underline it. Write a single-variable or compound inequality and solve. Write a sentence to explain your solution within the given context.

1) To maintain his current class average, Alonso needs to earn at least an 82 but no more than an 86 on his next math test. If each question is worth 2 points and there are 50 questions, what is the range in questions that Alonso can answer correctly and maintain his grade?

2) Santiago was using a cab service in the city. The cab service charged $2 per mile plus a $1 fee. Santiago could spend no more than $20 for his ride. What is the maximum distance he can travel in the cab?

3) Amir plans to walk more than 10,000 steps a day. He walks every morning and logs 4,700 steps. How many more steps does Amir need to walk to meet his goal?

Write a system of inequalities. Do not solve.

4) To maximize growing space for a vegetable garden, the Green family determined that the width of their garden box must be greater than 1.5 feet but no more than 3 feet. The maximum length of their garden box is 10.5 feet.

5) Camila was working on a science project for her chemistry class. Her teacher told her that if the chemical reaction took less than 2 minutes or more than 5 minutes, she would need to try the experiment again. She knows that each reaction will take at least zero minutes. Write a system of inequalities to represent when the experiment should be reattempted.

6) Brian's truck can hold 24 gallons of gas. He never lets his gas tank drop below a quarter full. On average, gas costs $3.00 in Brian's area, and he spends a maximum of $150 each month on gas.

7) A restaurant is placing a food order for chicken and fish. They must purchase at least 10 whole chickens and at least 6 whole fish. Whole chickens cost $8 and whole fish cost $20. The restaurant must spend less than $400 for this order. Write a system of inequalities where chicken is x and fish is y.

8) Graph the system of inequalities from problem 7. If the restaurant knows they must order 11 whole fish for a party, what are three possible chicken orders they can make and stay under budget?

9) A distribution warehouse needs to have at least 700 parking spaces for the employee vehicles and the company freight trucks. A car requires 180 square feet of space and a freight truck requires 450 square feet. The total parking lot space is 300,000 square feet. Write a system of inequalities.

10) Leonardo and Michelangelo started Turtle's Pizza. Each pepperoni pizza sells for $15 and costs $6 to make. Each veggie pizza sells for $12 and costs $7 to make. Expenses need to be below $840, while pizza sales need to be above $5,000. Write a system of inequalities for sales and expenses.

11) Julianna knits scarves and sweaters. Her goal is to have a combination of at least 20 scarves and sweaters to sell at the craft fair. A scarf takes 4 hours to knit and a sweater takes 10 hours to knit. Julianna has 150 hours that she can dedicate to her knitting. Write a system of inequalities where x represents scarves and y represents sweaters.

12) Graph the system of inequalities from problem 11. If Julianna can only knit 4 sweaters, will she still be able to reach her goal for the craft fair? Explain.

TARGETED REVIEW 18

◎ Targeted Review

In the Targeted Review, you will practice topics you have mastered in earlier lessons. Reviewing these concepts will help you be successful as you work through this unit.

Complete the problems on a separate sheet of paper.

Write the system of equations. Do not solve.

1) A coin purse contained $0.97 in pennies and nickels. Twenty-five coins were in the coin purse.

2) The Zobel family went to a local baseball game. At first, they ordered eight hot dogs and three drinks for $44.50. During the 7th inning, they ordered five hot dogs and seven drinks for $45.75.

3) Explain why graphing a system of linear equations in slope-intercept form may not always result in the exact solution.

4) Solve the system of equations.
$x - y = 5$
$10y + x = 3y + 3x$

5) Explain what the solution to a system of equations is in your own words.

6) Determine the equations of the vertical and horizontal lines passing through the point $\left(-6, \frac{2}{5}\right)$.

7) Determine the x-intercept for: $5x - 8y = 22$

8) Create a table of the relation $\{(-6, 8), (-3, 8), (0, 14), (3, 14)\}$ and name the domain and range for the relation.

Multiple Choice

_____ 9) Use the graph to find the difference between y and $2x$ for the system.

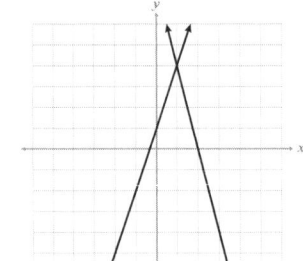

A) 7
B) 2
C) −2
D) −7

_____ 10) Determine the y-value for the system of equations.
$y = 2x + 1$
$y = x - 0.5$

A) −2
B) −1.5
C) −0.5
D) 1

11) Select all solutions that are true for the system of inequalities.
$y > -\frac{1}{2}x + 3$
$y \leq x$

☐ (6, 0)
☐ (3, 2)
☐ (2, 2)
☐ (10.5, −0.5)

12) Which of the following is true for the function? Select all that apply.
$f(x) = -\frac{5}{6}(x + 6)$

☐ $f(1) = 7$
☐ $f(2) = -\frac{20}{3}$
☐ $f(6) = -10$
☐ $f(-3) = -\frac{5}{2}$

Congratulations!

You have completed
Book A
of *Algebra 1: Principles of Secondary Mathematics!*

You are ready to continue the course in
Book B

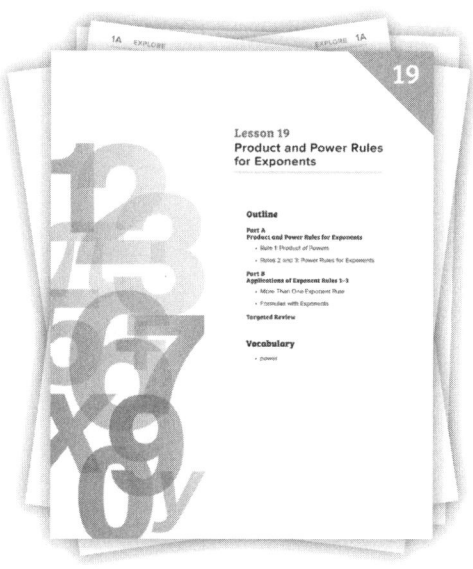

Index

A

absolute value, 3A, 29B
absolute value equations:
    graphing solutions to, 3A
    solving, 3A
    multi-step, 3B
    solutions to, 3A, 3B
    writing from a word problem, 3A
ac-**grouping method**, 23A, 23B, 24A
algebraic properties. *See* properties, algebraic
area models, 23A
arithmetic sequence, 14B
Associative Property, 1B
asymptote, 30A
axis of symmetry, 26A

B

bar graphs, 6A
bell curve, 6B
binomials, 20B, 21B, 22B
bivariate data, 13A, 13B
box organizer, 23B
break-even problems, 18A

C

clearing decimals, 2B
clearing fractions, 2B
coincident lines, 15A
coin problems, 17B
common difference, 14B
common ratio, 30A, 30B
Commutative Property, 1B
    with polynomials, 20B
compound interest, 30B
continuous function, 14A
coordinate plane, the, 7A, 12B, 15A, 30A
    ordered pairs, 7A
    quadrants, 7A
    x- and *y*-axis, 7A, 12B
correlations, 13A
correlation coefficient, 13A
cross product, 5A, 9A

D

data, 6A, 6B, 13A, 13B
    bivariate, 13A, 13B
    data sets, 6A, 6B, 13A, 13B
    quantitative vs. categorical, 6A
    normally distributed, 6B
decay (exponential functions), 30A, 30B
defining variables, 2A, 9A, 10B
difference of two squares, 22B, 24A, 29B
digit problems, 18A
dilation, 27B
dimensional analysis. *See* unit conversion
discrete function, 14A, 14B
Distributive Property, 1B
    with polynomials, 20B, 21A
    with proportions, 5A
domain, 7A, 11B, 14A, 26B, 30A
dot plots, 6A

E

elimination, 16B, 17A
Equality, Addition Property of, 1B
Equality, Multiplication Property of, 1B, 16B
equation of a line. *See* lines, equations of
exponential functions, 30A, 30B
exponents, 19A, 19B, 28A, 28B, 29B, 30A
exponent rules, 19A, 19B, 28A, 28B, 29A, 29B
    Rule 1 (product of powers), 19A
    Rule 2 (power of a power), 19A
    Rule 3 (power of a product), 19A, 28B
    Rule 4 (negative exponent), 28A
    Rule 5 (zero exponent), 28A
    Rule 6 (power of a quotient), 28B
    Rule 7 (quotient rule), 28B
    Rule 8 (fractional exponent), 29A
extending solutions (to a system), 17A
extraneous information, 10B
extraneous solution, 25A, 25B
extrapolation, 13B

INDEX

F

factoring polynomials, 21A, 21B, 22A, 23A, 23B, 24A, 24B, 25A
five-number summary, 6A
formulas with exponents, 19B
fractional coefficients, 2B
functions, 7A, 7B, 11B, 14A
    continuous, 14A
    discrete, 14A, 14B
    exponential, 30A, 30B
    function notation, 7B
    function rule, 7B
    mapping, 7A
    quadratic, 26A, 26B
    vertical line test (VLT), 7A, 11B

G

geometry formulas, 18A
graphing:
    absolute value equations, solutions to, 3A
    describing graphs, 9A
    exponential functions, 30A
    inequalities, solutions to, 4A
    linear equations (lines), 8A, 9A, 9B, 11A, 11B, 15A
    linear inequalities, 15B
    number lines, solutions on, 1A
    parabolas, 26A, 26B, 27A, 27B
    quadratic functions, 26A, 26B, 27B
    quadratic inequalities, 27A
    scatter plots, 13A
    sketching graphs, 9A
    systems of equations, 15A, 17A
    systems of linear inequalities, 15B
    using technology, 13A, 13B, 26B, 27A
greatest common factor (GCF), 11A, 21A, 21B, 24A
growth (exponential functions), 30A, 30B
growth/decay formula, 30B

H

histograms, 6A
    estimating the mean from, 6A
horizontal lines, 11B, 12B
how much or how many problems, 17B

I

Identity Property, Additive, 1B
Identity Property, Multiplicative, 1B
index (radicals), 29A
inequalities, 4A, 15B
    absolute value, 4B
    compound, 4B
    graphing solutions to, 4A
    linear, 15B, 18B
    multiplying by a negative, 4A
    quadratic, 27A
    solving, 4A
    symbols to, 4A, 15B, 27A
integers, 1A, 11A, 28B
intercepts, x- and y-, 8A, 8B, 11A
interpolation, 13B
intervals, 14A
interval notation, 14A
inverse operations, 2A
Inverse Property, Additive, 1B, 16B
Inverse Property, Multiplicative, 1B
irrational numbers, 1A
    classification of, 1A
    operations with, 1A

L

Least Common Denominator (LCD), 2B
linear combinations, 16B
linear inequalities, 15B, 18B
line of best fit, 13B
lines, equations of, 8B
    applications of, 10B
    comparing forms of, 11A
    continuous vs. discrete, 14A
    horizontal, 11B
    introduction to, 7A, 7B, 8A, 8B
    line of best fit, 13B
    parallel, 12A
    parent function, 8B
    perpendicular, 12B
    point-slope form, 9B
    slope-intercept form, 9B, 10A
    standard form, 11A
    vertical, 11B

M

measures of center, 6A
    mean, 6A
    median, 6A
    mode, 6A
measures of spread, 6A
    range, 6A
    range, interquartile, 6A
    standard deviation, 6A, 6B
midpoint, 3A
modeling factors, 23A
monomials, 20B, 21A

N

natural numbers, 1A
normal distribution, 6B
numbers:
    classifications, 1A
    ordering on a number line, 1A
    placement on real number diagram, 1A

O

ordered pairs, 7A, 8A,
    in words, 9A
    with continuous and discrete functions, 14A
    with linear equations, 9A, 9B, 10A, 10B
    with lines of best fit, 13B
    with quadratic functions, 26A
    with scatter plots, 13A
    with slope, 9A
    with systems of equations, 15A, 16A
origin, the, 7A, 8B, 27A
outliers, 6B, 13B
    formula for, 6B

P

parabolas, 26A, 26B, 27A, 27B
parallel lines, 12A, 15A
parent functions, 8B, 27B
perfect squares, 22B
perfect square trinomials, 22B, 24A
perpendicular lines, 12B
PIE Method, 2A
Plan, Implement, Explain Method, 2A
point-slope form, 9B, 10A, 12B, 14A

polynomials:
    adding and subtracting, 20A
    classifying, 20A
    difference of two squares, 22B, 24A, 29B
    factoring, 21A, 21B, 22A, 23A, 23B, 24A, 24B, 25A
    perfect square tirnomials, 22B, 24A
    multiplying, 20B
    rational expressions, 28A, 28B
    sign patterns, 22A, 23A, 24A
    solving for an unknown coefficient, 20A, 20B
    solving polynomial equations, 24B, 25A, 25B, 26A, 26B
    solving quadratic equations, 24B, 25A, 25B, 26A, 26B
properties, algebraic:
    Associative Property, 1B
    Commutative Property, 1B
    Distributive Property, 1B
    Equality, Addition Property of, 1B
    Equality, Multiplication Property of, 1B
    Identity Property, Additive, 1B
    Identity Property, Multiplicative, 1B
    Inverse Property, Additive, 1B
    Inverse Property, Multiplicative, 1B
    Reflexive Property, 1B
    Substitution Property, 1B
    Symmetric Property, 1B
    Zero-Product Property, 1B
Property of Symmetry, 1B
proportions, 5A

Q

quadrants. *See* coordinate plane, the
quadratic equations, 24B, 25A, 25B, 26A, 26B
quadratic functions, 26A, 26B, 27B
quadratic inequalities, 27A
quartiles, 6A

INDEX

R

radicals:
- equations, 29B
- expressions, 29A
  - addition and subtraction of, 29A
- index, 29A
- radicand, 29A, 29B
- sign/symbol, 29A
- terms, 29A
- vinculum, 29A

radicand, 29A
range, 7A, 11B, 14A, 26B, 30A
range, interquartile, 6A
rate of change, 9A, 9B, 10B
ratios, 5A, 9A
- equivalent, 5A
- simpliflying, 5A

rational expressions, 28A, 28B
rational numbers, 1A
- classification of, 1A
- operations with, 1A

real numbers, 1A
real number system, 1A
real number diagram, 1A
reflection, 27B
Reflexive Property, 1B
regression equation, 13B
relations, 7A, 13A
rise over run. *See* slope
roots, 26A, 29A, 29B

S

scatter plot, 13A, 13B
sequence, 14B
shorthand, mathematical, 1A
sign patterns, 22A, 23A, 24A
simple interest, 30B
slope, 8A
- calculating, 9A
- introduction to, 8A
- formula, 9A, 11A
- graphing, 8A
- linear parent function, of the, 8B
- rate of change, 9A, 10B
- when comparing equations, 11A
- with horizontal and vertical lines, 11B
- with parallel lines, 12A
- with perpendicular lines, 12B
- with point-slope form, 9B
- with slope-intercept form, 9B, 10A
- with systems of equations, 15A

slope-intercept form, 9B, 10A, 12A, 12B, 14A
slope formula, 9A
solutions:
- to absolute value equations, 3A, 3B
- to equations, 2A, 2B
  - all real numbers, 2B;
  - one solution, 2A, 2B;
  - no solution, 2B;
  - quadratic, 24B, 25A, 26A, 26B
  - radical, 29B
- to inequalities, 4A
  - absolute value inequalities, 4B;
  - compound inequalities 4B
  - linear inequalities, 15B, 18B
  - quadratic inequalities, 27A
- to systems of equations, 15A, 16A, 16B, 17A, 17B, 18A
- to systems of linear inequalities, 15B, 18B

solving equations:
- containing an absolute value, 3A
- containing decimals, 2B
- containing exponents, 29B
- containing fractions, 2B
- from a word problem, 2A
- multi-step, 2A
- multi-step containing absolute value, 3B
- polynomial, 24B, 25A, 25B, 26A
- quadratic, 24B, 25A, 25B, 26A
- radical, 29B
- solving systems of equations, 15A, 16A, 16B, 17A, 17B, 18A
- with more than one variable, 2A

standard deviation, 6A, 6B
- 68-95-99.7 Rule, 6B

standard form of a line, 11A, 14A, 16B
standard form of a polynomial, 20A, 23A, 26A
substitution, 16A, 17A
Substitution Property, 1B
Symmetric Property, 1B

systems of equations, 15A, 17A, 17B, 18A
    solving using elimination, 16B
    solving using graphing, 15A
    solving using substitution, 16A
systems of linear inequalities, 15B, 18B

T

terms (within expressions), 1B
transformations, 27B
translation, 8B, 12A, 27B
trend line. *See* line of best fit
trinomials, 20B, 22B

U

unit conversions, 5B
    compound, 5B
unit multipliers, 5B

V

variables:
    defining, 2A, 9A, 10B
    equations with more than one, 2A
    exponents, that are, 29B
    independent vs. dependent, 7B, 9A, 13A
vertex, 26A, 26B, 27B
vertex form, 27B
vertical lines, 7A, 11B, 12B
vertical line test (VLT), 7A, 11B
vinculum, 29A

W

whole numbers, 1A, 11A
wind and water problems, 17B

X

x-intercept, 8A, 8B, 11A, 24B, 26A
X organizer, 23A

Y

y-intercept, 8A
    introduction to, 8A
    meaning in wording problems, 10B
    when comparing equations, 11A
    with exponential functions, 30A
    with horizontal and vertical lines, 11B
    with parallel lines, 12A
    with perpendicular lines, 12B
    with quadratic functions, 26A
    with quadratic inequalities, 27A
    with slope-intercept form, 9B, 10A
    with standard form, 11A
    with the linear parent function, 8B

Z

Zero-Product Property, 1B
    with polynomials, 24B

REFERENCES

The use or mention of a supplemental resource does not imply endorsement by the author and publisher, nor does it imply that said resource endorses this book.

The author and publisher specifically disclaim all responsibility and accept no liability for any loss, injury or risk, personal or otherwise, resulting from or arising out of the use and application of this book and its contents, including without limitation projects and resources mentioned herein.

Please note that online resources that were available at the time of publishing may have been updated or may no longer be available.

Desmos® is a registered trademark of Desmos Incorporated in the United States and/or other countries.